UNDERSTANDING ORIGINS

BOSTON STUDIES IN THE PHILOSOPHY OF SCIENCE



VOLUME 130

# UNDERSTANDING ORIGINS

## *Contemporary Views on the Origin of Life, Mind and Society*

*Edited by*

FRANCISCO J. VARELA AND JEAN-PIERRE DUPUY

*CREA, École Polytechnique, Paris, France*

KLUWER ACADEMIC PUBLISHERS

DORDRECHT / BOSTON / LONDON

**Library of Congress Cataloging-in-Publication Data**

Understanding origins : contemporary views on the origin of life, mind, and society / edited by Francisco J. Varela and Jean-Pierre Dupuy.
p. cm. -- (Boston studies in the philosophy of science ; v. 130)
"The material comes from an international meeting in September 13-16, 1987 at Stanford University"--Introd.
Includes index.
ISBN 0-7923-1251-1 (alk. paper)
1. Life--Origin--Congresses. 2. Social structure--Origin--Congresses. 3. Cognition--Origin--Congresses. 4. Money--Origin--Congresses. 5. Money, Primitive--Congresses. I. Varela, Francisco J., 1945- . II. Dupuy, Jean-Pierre, 1941- III. Series.
QH325.U53 1991
577--dc20 91-15433

ISBN 0-7923-1251-1

---

Published by Kluwer Academic Publishers,
P.O. Box 17, 3300 AA Dordrecht, The Netherlands.

Kluwer Academic Publishers incorporates the publishing programmes of D. Reidel, Martinus Nijhoff, Dr W. Junk and MTP Press.

Sold and distributed in the U.S.A. and Canada
by Kluwer Academic Publishers,
101 Philip Drive, Norwell, MA 02061, U.S.A.

In all other countries, sold and distributed
by Kluwer Academic Publishers Group,
P.O. Box 322, 3300 AH Dordrecht, The Netherlands.

*Printed on acid-free paper*



Printed in The Netherlands

# TABLE OF CONTENTS

## PART IV: PERCEPTION AND THE ORIGIN OF COGNITION

# PREFACE

The main intention of this book is to bring together contributions from biology, cognitive science, and the humanities for a joint exploration of some of the main contemporary notions dealing with the understanding of origins in life, mind and society.

The question of origin is inseparable from a web of hypotheses that both shape and explain us. Although origin invites examination, it always seems to elude our grasp. Notions have always been produced to interpret the genesis of life, mind, and the social order, and these notions have all remained unstable in the face of theoretical and empirical challenges. In any given period, the central ideas on origin have had a mutual resonance frequently overlooked by specialists engaged in their own particular fields.

As a consequence, this book should be of interest to a wide audience. In particular, for all those engaged in the social sciences and the philosophy of science, it is unique document, since bridges to the natural sciences in a mutually illuminating way are hard to find. Whether as a primary source or as inspirational reading, we feel this book has a place in every library.

The material comes from an international meeting held in September 13—16, 1987 at Stanford University, organized by F. Varela and J.-P. Dupuy at the request of the Program of Interdisciplinary Research of Stanford University. We are grateful to René Girard, the Program Director, for making it possible with the help of the Mellon Foundation. Our thanks also to Laurence Helleu for her skillful editorial help in the preparation of this book for publication.

In preparation for the Symposium, Andrew McKenna, André Orléan, Stuart Kauffman, Thomas Bever, and Francisco Varela were asked to prepare position papers which were circulated in advance to invited discussants who presented their comments during the meeting itself. The full program and list of participants is included in an Appendix. This book contains revised and updated versions of the position papers, a selection from discussants' presentations, and special lectures given by René Girard and Umberto Eco. The exception is the section on the



*Francisco J. Varela and Jean-Pierre Dupuy (eds), Understanding Origins,* vii—viii.

'Origin of Language' and Bever's paper, which, for various reasons, could not be included in this book.

The Introduction that follows provides the reader with a more detailed idea of the intellectual motivations behind this meeting, and some of the unifying threads running through the various contributions.

JEAN-PIERRE DUPUY AND FRANCISCO J. VARELA

# UNDERSTANDING ORIGINS: AN INTRODUCTION

We wish to start with the following observation: the humanities and the 'hard' sciences (here meaning especially biology and a good part of the cognitive sciences) differ considerably in their ambitions concerning the 'big questions'. The hard sciences are more daring than ever in proposing how the cosmos formed and life originated, how species evolved and the destiny of it all. In contrast, for the humanities it has been a time of dispersion, of fragmentation, of a dissemination which resists any attempt at integration on a grand scale. The time of the 'big theories' seems to have been left far behind.

It was not always so, and we can point to 1939, the year of Freud's death, as the turning point. Until then, the tradition of religious anthropology did not hesitate to postulate an origin to human cultures, a question which was inseparable from that of the origin of the religious. The central figures in this tradition — Fustel de Coulanges, Robertson Smith, Frazer, Durkheim, Hocart, and Freud — shared the hypothesis that every human institution is grounded in the religious, or, to be more precise, in ritual. A myriad of factors could be invoked to explain why this tradition disappeared quickly after World War II, and the subsequent schools considered it as plainly irrelevant. We may think, for example, of the structuralist movement in France and its postulation that structures are 'always already there'. Or the more positivistic bent of the Anglo-Saxon anthropology, personified for example by Evans-Pritchard. But more importantly we want to reflect on the general impact on the humanities of the work of Nietzsche and Heidegger, continued in the school of Jacques Derrida centered on the so-called 'deconstruction of Western metaphysics'.

The main question of metaphysics is, according to Heidegger, that of the origin of entities: why is there something rather than nothing? Faced with this question, Western metaphysics answers with the Leibnizian principle of sufficient reason: every effect has a cause, this arises from another cause, and so on, until one arrives at the first cause, self-sufficient, full and cause of itself, that is, God. This is the onto-theological argument: to the why-question one answers with the postulate of a



*Francisco J. Varela and Jean-Pierre Dupuy (eds), Understanding Origins*, 1—25.

fundamental entity, which evacuates the ontological difference between Being and entities. The mystery of there-is thus disappears under the weight of the answer.

Heidegger, in contrast, proposes no response for he begins by deconstructing the very question: it is necessary not to search for a cause, but to let the mystery unfold fully; the answer is a question mark. "La rose est sans pourquoi" (Rimbaud), she has no reason or cause. Derridian deconstruction picks up along the same Heideggerian theme. It will concentrate on those philosophical creations such as Nature, Language, Reason, Origin, Meaning, Truth, and Subject, which appear as full, self-sufficient and the cause of themselves. The point is to deconstruct the claim of Logos to affirm itself as complete and self-sufficient, the ambition of philosophy to have immediate access to pure truth (*aletheia*), the illusion of mastery on the part of the human subject who puts himself in the place of God. In its "deconstruction of Western metaphysics", the intellectual enterprise launched by Derrida and his numerous followers in the wake of Heidegger, systematically debunks the Concept which, like the self-centered bourgeois ridiculed by Marx in *The Holy Family* "swells up to the point of taking himself for an atom, that is to say a being devoid of any relation, sufficient unto himself, without needs, absolutely complete, in a state of complete felicity".

The major tool used by Deconstruction is what Derrida calls the *logic of the supplement*. As René Girard says in the text that follows, this logic "reflects [the] general human inaptitude to self-centeredness, [the] failure of individual and collective narcissim, and the resulting fear of and fascination with otherness". This logic is the one through which every philosophical text deconstructs itself. Every time that a term appears in a theoretical text which beckons a Logos, a Concept, as self-sufficient, a vicious circle sets in, which undermines this pretension to autonomy from within. This happens because another term, supposed to be secondary and subordinated, and which should be nothing other than a derivation or complication of the primary Concept (for instance: culture, writing, form, etc.), appears as indispensable to the constitution of the latter. The origin appears as full and pure but, without the supplement which nevertheless follows from it, it would lose all consistency. Thus the secondary term appears at the same time as perfectly dispensable and perfectly indispensable. Even the most apparently perfect totality suffers inescapably from a constitutive lack.

The logic of the supplement, then, can be depicted in the form of a circular causality unifying two terms in spite of the fact that one claims to be hierarchically superior to the other, as depicted in Fig. 1.

The deconstruction of a hierarchical opposition, it should be remarked, is not the same as its simple removal. The hierarchical dimension must remain present one way or another. And neither does deconstruction consist in simply inverting the hierarchical opposition, in permuting its superior and inferior terms. Take the example, especially important for Derrida, of the hierarchical opposition between philosophy and writing. Philosophy devalues writing precisely because it is written! Writing constitutes a threat to philosophy in the same manner that money does to economics, because it is an obstacle, a barrier in the way of access to meaning and value. Since the ideal of philosophy is to reach the truth without mediation it must therefore deny the only means it has of expressing itself: writing. Bluntly put, philosophy *writes* **W**:

**W**: '*This is not writing*'

the obvious form of a self-referential paradox. We will return to this issue of self-reference frequently.

There is little interchange between the activity of deconstruction and scientific production. The deconstructors can well savor the illusion that they hold in their hands in fact a non-logic, since not only does it

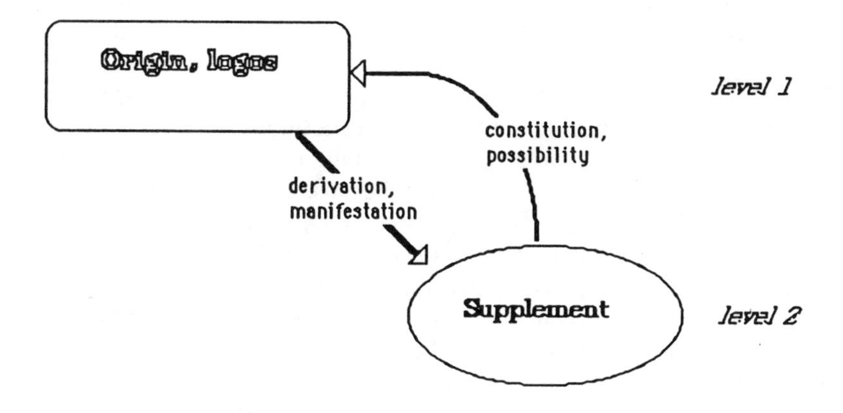


Fig. 1.

seem to elude *prima facie* scientific formalization, but it also undermines any postulations about an origin. Just take any scientific narrative dealing with the origin of some domain — life, meaning, language, money — and one could then bet that this discourse would self-deconstruct according to the logic of the supplement.

Let us consider an important example which well matches this expectation: molecular and cellular biology. It seemed that this discipline is a model success in the reduction of life to macromolecular chemistry, mainly through the discovery of the genetic code and the notion of a cellular programming which is supposed to stand at the base of all development as it (literally) writes the organism as it unfolds in its ontogeny. However, after an initial phase of enchantment with the idea, it has become clear — and the molecular biologists were the first to point this out — that if one takes the notion of a genetic program literally one falls into a strange loop: one has a program that needs its own product in order to be executed. In fact, every step of DNA maintenance and transcription is mediated by proteins, which are precisely what is encoded. To carry on the program it must already have been executed! Thus we can depict the situation as shown in Fig. 2.

Now, the practicing biologist does not lose sleep over this fact. For him the paradox is resolved since every cell is already derived from another cell, and thus an individual ontogeny already starts from the

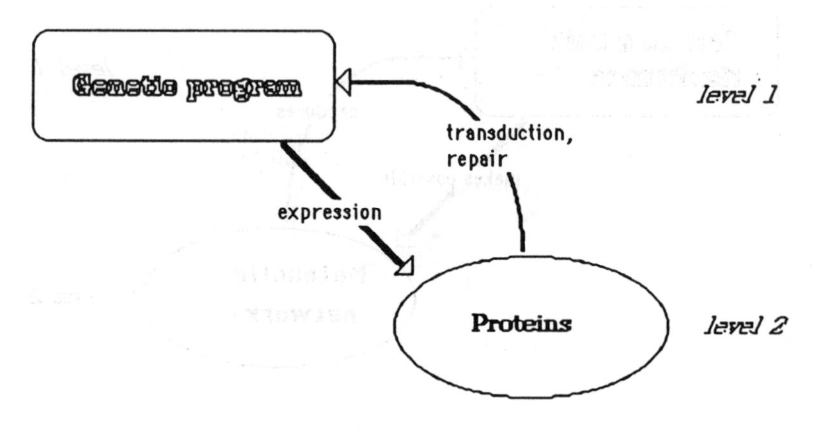


Fig. 2.

result of the mother's fertilized egg. From a more theoretical viewpoint, the issue is more thorny since at some point one must reach the origin of life and initiate the chain of autonomous individuals. These efforts to ground the autonomy of the living have taken precisely the form of an apparently paradoxical loop or *autopoiesis*:[1] the logic of the cell is that of self-production by a circular determination between its boundaries and its dynamics, which both produces the boundaries and is made possible by them, as indicated in Fig. 3.

For theoreticians this presents a profound interest: the observation that under certain conditions an assembly of components can link up in some intricate, circular causality — or to be more technical, by operational closure:[2] — so that there is an emergent new level which is neither reducible to the sum of its components nor is separate from its products. Whence the apparently paradoxical loop between two levels which appear to have, at first glance, a hierarchical relation (e.g. the cell as a unit vs. its chemical dynamics), and yet are hopelessly intermixed. The form of this logic is then that of two levels which must be kept distinct, and yet which are undeniably intertwined.

This kind of active self-reference is at the heart of the matter we wish to address in this book. In other words, it is one of our main contentions that the self-referential qualities of operative processes at various levels share common features, the most important one being

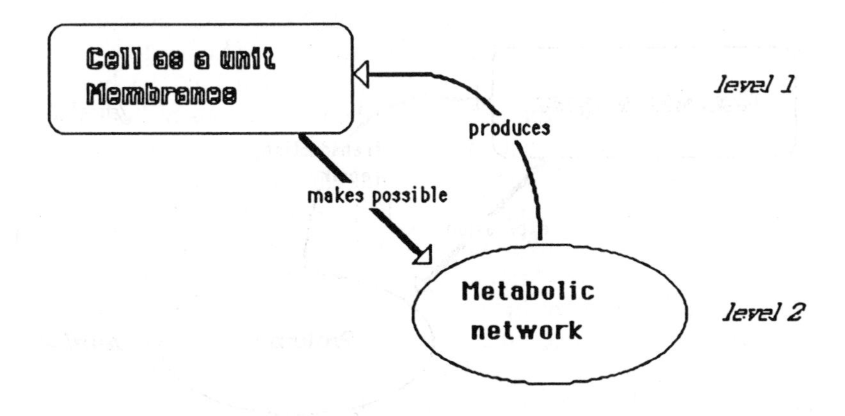


Fig. 3.

the constitution of a unitary entity (be it a cell system, a language, or a monetary system) that seems exterior to its components, and yet which is endogenously engendered by the intertwining of those very components. In this sense what may appear as exogenous (the Value, the Other) may be perfectly consistent with its being endogenously constituted if one looks at the entire inclusive logic of the situation.

Now, it is remarkable that those theoreticians of natural autonomous systems who have attempted to deal with the original grounding of life end up with a logic which mirrors the Derridian logic of the supplement, the main weapon to arrive precisely at the negation of all pretension to autonomous grounding and self-sufficiency! One of the main purposes of the meeting, and a main guiding line in this book, has been to try and break down the academic walls that have kept these two distinct modes of work separate, to confront them, and to search for further clarification. Our approach to this task has been to take specific notions of origin in life, mind and society as case studies. Let us now pause before returning to the more general principles we have addressed in this section, in order to consider the basic outlines of the case studies included here.

It seemed inevitable that we should devote a session during the meeting to the question of the origin of social order as seen from the vantage point of the fundamental anthropology of René Girard, whose special lecture is included here. We also asked Andrew McKenna to discuss the interface between Girard and Derrida, and have included the further commentaries by Paul Dumouchel and Paisley Livingston.

Girard's is indeed the only extant approach that links back to the classical religious anthropology, defying the interdictions of both structuralism and its deconstruction. It does not hesitate to confront the question of the origin of the sacred and, through it, the origin of all social and cultural institutions. As we shall see, however, the core of the logic used by Girard is none other than that of deconstruction and of the theory of autonomous systems.

We cannot summarize here the vast and complex work of Girard, which has unfolded over the years by various presentations which constitute progressive stages.[3] Two main foundational hypotheses are important for us here, the second being, in principle, derivable from the first. First and foremost the hypothesis of mimetism: men imitate each other in their desire; we never desire anything other than what the other desires. This mimesis of appropriation ends up inevitably in conflict and violence: the other mutates from a model, to automatically become a rival, an obstacle.

This leads to the second hypothesis of victimization: all primitive societies would supposedly have lived a primordial and foundational event: at the climax of the war of all against all unleashed by the mimesis of appropriation, there was a polarization of all exacerbated violence focused on one arbitrary member of the group. These sacrificial exclusions, lived in the ignorance of that mechanism, would be the source of the sacred, of culture, of all human institutions.

Now it is certainly possible to read the Girardian thesis from the vantage point of deconstruction — which is not the same thing as deconstructing Girard with the tools of deconstructivism. This is what McKenna does in his text. "Girard's originary scenario can accommodate a post-structuralist critique of origins because it posits that origin and *différance*[4] are one, that representation is the by-product of an originary *différance*, of mimetic desire, rather than the representation of an originary presence, of an origin of any kind." One can argue with McKenna that in *both* Girard and Derrida, "in the beginning is imitation, not origin". This is true in a sense for the mimetic mechanism: the original situation cannot be one of a subject *A* who imitates the desire of another subject *B*, who desires autonomously, since mimesis is universal. One is forced then to assume a double imitation, where *A* imitates *B* and *B* imitates *A*. Objects can flash out from this mechanism according to the logic of self-fulfilling prophecies. *A* believes that *B* desires *O*, makes the first step towards *O*, thus signifying to *B* that *O* is desirable. When *B*, in turn, manifests his desire, *A* has the proof that he was not wrong. The object *O* is not an origin, since it is in fact an effect, a supplement; at the origin there is only repetition.

A similar reasoning applies to the mechanism of the victim. The sacrificed victim — made sacred — pays for the others, is a scapegoat. But the use of this expression reveals a certain knowledge about the victimization mechanism that the collectivity of persecutors cannot have, for, if they did, that would mean the end of the said mechanism. McKenna points out that "the victim is always already a substitute, a signifier", and adds "a mark, in Derridian terms, of a deferral". Collective sacrifice, in fact, calms violence through violence, and is always a deferral of violence. Thus McKenna concludes: "Nothing Derrida advances by way of his critique of origins . . . is proof against Girard's anthropological hypothesis of human origins as rooted in the dynamics of mimetic desire."

Another choice would be to read Derrida in the light of Girard. This is what Girard himself does in his text here. He shows how the "theories of origin of non-scientific culture", that is, the foundational myths, are at least as much if not more than so-called Western metaphysics, embedded in the logic of the supplement. Girard explains this by using his own anthropological hypothesis: if the myths about the origin are structured around a paradoxical logic it is because they tell a story that *actually* happened, but which was lived in ignorance (*méconnaissance*). A society becomes unitary by way of an exclusion: that of the victim which thereby becomes sacred; without such an exclusion it would not exist. The myth expresses at the same time the internal and external nature, the indispensable and dispensable character, and the infinitely good and infinitely evil, of the sacralized victim. The latter *is* the supplement. The logic of the supplement is the logic of a narrative which tells of a real event with distortions which are not random, but well-defined and regulated. This supplement is at the heart of all religious thinking and, because the religious subsists in the philosophical, philosophy is also undermined by the supplement. "The logic of the supplement must be mythical first and philosophical second."

It is this realist epistemology that discussants Dumouchel and Livingston address in their texts. Dumouchel remarks that Girard's is a morphogenetic theory: it can account for the emergence of new forms. The mimetic mechanisms can produce new complexifications, the simple can give birth to the complex. The Girardian origin does not contain what will come out of it; it is not an essence. The Derridian critique has no effect on Girard's theory since, as Derrida remarks, one cannot criticize metaphysics except with metaphysical concepts. And as Dumouchel says: "Implicit in Derrida is a concept of origin which cannot accommodate his own criticism of origin." The Girardian theory of the origin of culture allows one to posit, and to conceive precisely, an origin which is not an essence.

In the same direction, Livingston also asserts that it is possible and desirable to unclutter Girardian thinking of all remaining metaphysics. To accomplish this it is necessary to plunge Girard into a 'naturalistic framework', which is, according to Livingston, what Girard himself wants to do: "Girard seeks to avoid the kind of circularity a deconstructionist expects to find in any theory of the origin of representation." To make this possible, one must root human mimesis in animal mimesis, to consider that "this mimesis has an origin: it has evolved or emerged as a natural reality within the natural universe." This entails the existence of a purely instinctual mimesis, "an immediate and mechanical form of mimicry." One can liberate original mimesis of all representational and intentional aspects. In the origin there would not be mimetic 'desire' but simply a mimesis of appropriation. In this sense Girard can account for this world of representations which is culture from a hypothesis where representations play no role.

If it is possible to read Girardian theory on the one hand in the sense of deconstruction, and on the other hand under a realist/positivist epistemology, it is because of a specific aspect of its configuration, namely that the origin it postulates is a supposedly real event, but which can only be produced because its reality is *méconnue*, misread by those who are the actors. Theory itself explains why the origin is real but inaccessible, being one of those strange objects which do not exist except in ignorance: see Fig. 4.

It is our contention, then, that Girardian theory, and other theories of origin, as we shall see shortly, require a more adequate philosophical framework. This would be an epistemology that neither seeks an ultimate, real grounding in the style of the hard sciences, nor does it satisfy itself in the nihilism of a permanent deconstruction. What is needed is some middle way, a meta-position which does not need

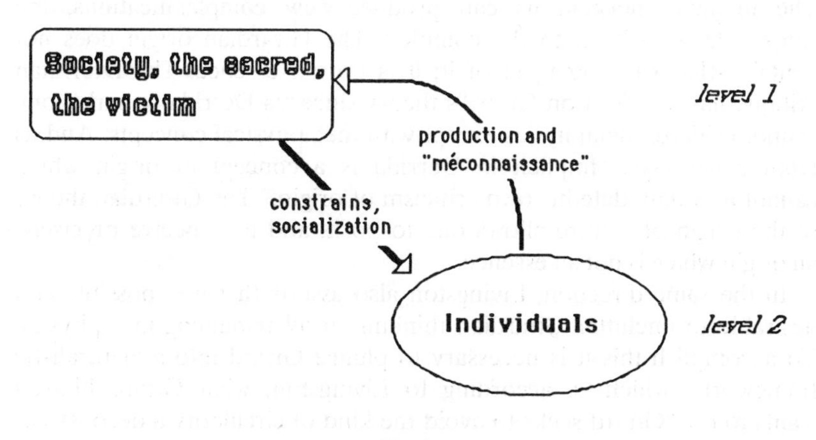


Fig. 4.

ultimate grounding while at the same time seeks an original mechanism for grounding. The apparent contradiction of this position is resolved, of course, by demanding that the mechanism of origin partake of the morphogenetic and paradoxical qualities that we have outlined so far, as exemplified in the two apparently unrelated cases of cellular organization and Girardian mimesis. This is the core of the interpretation we propose for a fresh look at the question of origin. But we are getting ahead of ourselves. Let us now return to consider the content of the other texts in the book.

The session on the origin of money was chosen precisely because it illuminates in an exemplary manner the core issues we have been discussing. In fact, the question of what gives the monetary token its value has been the subject of the most contradictory interpretations. Within this spectrum we find again the two extremes of the excavation of a real, absolute, grounding original event; and on the other the deconstruction into complete nihilism. The reading by Orléan of modern economic theories points precisely in the direction of some middle way of auto-foundation.

For a long time the economics of money has been dominated by the metallist tradition, which holds "that the foundational acceptance of money as medium of exchange should be understood as a belief in the intrinsic value possessed by money itself." There is some good which has a 'true' value, usually referring to gold, and money owes its value to the fact that it contains or that it is at least backed by this real value of gold.

Few people today hold this position in face of the evidence that paper money — still called *fiat* money — still has value even if it has no intrinsic value and it is non-convertible. It is this evidence which explains why both structuralism and deconstructivism have abundantly used monetary metaphors for deconstructing the metaphysical conception of a sign as corresponding to a transcendental carrier of significance. These modern and postmodern demystifiers are happy to point to paper money as an example of signification: a mere 'autonomous' sign, without a referent, a sign of a sign, a copy of a copy, a simulacrum. The central treasure which should consecrate the value of money is empty, and this must be proclaimed loud and clear.

Orléan shows how it is possible to find another alternative in this sterile opposition. The origin of money resides neither in some ultimate grounding, nor in an inaccessible time since always deferred, but in a *mechanism* of self-organization and of auto-exteriorization. This bootstrapping mechanism is the same one through which society projects itself as existing outside of itself, as it were.

A great originality in the approach of Orléan is his focus on the continuity between primitive and modern money, since "both are expressions of the social whole as a separate entity." He thus reformulates the analysis of anthropologists such as Barraud, de Coppet, Itéanu and Jamous when they say: "there is no money in the absence of a transcendent order that gives it the quality of being a materialization of the social totality."

This common reality makes it possible to distinguish modern from primitive money. For archaic societies it is the sacred which realizes and 'materializes' the movement through which society exteriorises itself (and Orléan makes here a link with Girard's anthropology). There is a wealth of anthropological observations on this, and Orléan describes some of them, showing primitive money to be indissociably linked to the sacred. Modern societies do not have access to the same mechanism of exteriorization, and we are not surprised to see that ultra-individualist monetary theories, because they deny the transcendence of the social in regard to individuals, are led to deny the reality of money itself!

Orléan favors a particular monetary view, that of Keynes, because he thinks that it makes it clear how a process of auto-externalization is possible at all when considered in an individualist framework. Money and individualistic values can then be reconciled. The reality of money takes the form of conventions, always partly contingent and without grounding in some ultimate Reason or Nature, but without which society would not exist at all. This social activity leading to its viewing a part of its own doing as exterior to itself — money acquiring a value which seems to be given elsewhere — thus introduces a certain opacity which would be absent if some ultimate grounding was forthcoming. The nature of this process of conventions then appears as another instance of a highly interactive network of agents who find mutual satisfaction by their simultaneous acting. This endogenous solution acquires, in the eyes of the actors, an external value, thus creating precisely the logic of the supplement we have already encountered, as depicted in Fig. 5.

Jean-Joseph Goux points out the persistence of some features of

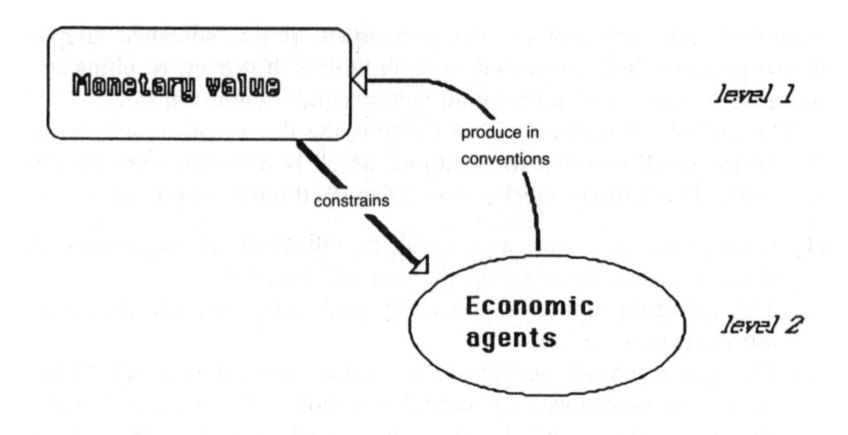


Fig. 5.

primitive money into modern money, starting with its sacrificial dimension: there is no standard of value or meaning without a kind of radical exclusion akin to a sacrifice, even in the case of modern money. To be sure, as the Greek case already shows, modernity introduces a fundamental rupture which can account at the same time for the birth of modern money, the rise of individualism, and the emergence of the ideal of social autonomy or self-institution. The Self as viewed by the moderns (and already by the Greeks) implies a dimension of transcendence (hence of sacrifice), but in the form of self-transcendence (self-sacrifice). Individualism tends to internalize this movement more and more within the individual subject himself. In the monetary domain the best illustration of this paradoxical de-transcendentalization is the circulation of the formerly transcendent standard of value as a medium of exchange, a means of circulation in the profane market.

Evolution and the question of how the living comes to be as we find it was, obviously, a central theme of the Conference as raised by Stuart Kauffman's main paper, and retaken by various commentators: Brian Goodwin, Susan Oyama, John Dupré, and Daniel Brooks. It must be said at the outset that the choice of participants for this session was quite explicitly from the critical side of current evolutionary biology. In fact, although the notion of adaptation is the centerpiece of much of recent evolutionary biology, many critiques of this so-called adaptationist program have appeared in recent years, however, resulting in a full scale revision of what was, until quite recently, a uniform view.[5]

The orthodoxy under revision today is the theory of organic evolution in its neo-Darwinian formulation, which is relatively easy to state succinctly. This heritage can be summarized in three basic points:

(1) Evolution occurs as a gradual modification of organisms by descent; that is, there is reproduction with heredity.
(2) This hereditary material constantly undergoes diversification (mutation, recombination).
(3) There is a central mechanism to explain *how* these modifications occur: the mechanism of natural selection. This mechanism operates by selecting the designs (phenotypes) which cope with the current environment more efficiently.

This classical Darwinism became neo-Darwinism during the 1930s as a result of the so-called 'modern synthesis' between the Darwinian ideas based on zoology, botany and systematics, on the one hand, and the increasing knowledge of cellular and population genetics on the other. This synthesis established the basic view that modifications occur by small changes in organismic traits specified by heritable units, the genes. The genetic makeup responsible for the ensemble of traits leads to differential reproduction rates, hence to changes in the genetic makeup of an animal population over generations. Evolution is simply the totality of these genetic changes in interbreeding populations. The pace and tempo of evolution are measured by the changes in the fitness of genes; thus it is possible to give a quantitative basis for the visible adaptation of animals to the environments in which they live. These concepts are, of course, ones with which we are all familiar. But we need to clarify them one step further in order to do justice to their multiple scientific roles.

Consider the concept of adaptation. The most intuitive sense of adaptation is that it is some form of *design* or construction that matches optimally (or at least very well) some physical situation. For example, the fins of fishes are well suited for an aquatic environment, whereas the ungulate hoof is well suited for running on the prairies. Although this conception of adaptation is quite popular, most professional evolutionary theorists do not construe adaptation in this way. Instead, adaptation has come to refer specifically to the *process* that is linked to reproduction and survival, i.e. to adapting. This process is — or so one supposes — what accounts for the apparent degree of adaptational design observed in nature.

To make this idea of adapting do theoretical work, however, we need some way to analyze the adaptedness of organisms. This is where the notion of *fitness* comes in. From the vantage point of adaptedness, the task of evolution consists in finding heritable strategies, i.e. sets of interrelated genes that will be more or less capable of contributing to differential reproduction. When a gene changes so as to improve in this task, it improves its fitness. This idea of fitness is often formulated as a measure of abundance. It is usually taken as a measure of individual abundance, i.e. as a measure of the surplus offspring achieved, but it can also be construed as a measure of population abundance, i.e. as the effect of genes on the rate of growth of a population.

It has become increasingly clear, however, that this way of measuring fitness as abundance has a number of conceptual and empirical difficulties. First of all, in most animal groups reproductive success depends on the sexual encounters with other individuals. Second, since the effects of any given gene are always intertwined with a multitude of other genes, it is not always possible to differentiate the effects of individual genes. Third, the milieu in which the genes are supposed to express themselves is enormously varied and time-dependent. Finally, this milieu must be viewed in the context of the entire life cycle and ecology of an animal. Fitness can also be taken as a measure of *persistence*. Here fitness measures the probability of reproductive permanence over time. What is optimized is not the amount of offspring, but the probability of extinction. Clearly this approach is more sensitive to long-term effects, and so it is an improvement over the more narrow view of fitness as abundance. By the same token, however, it poses formidable problems at the level of measurement. Virtually all of these points are raised in the following texts.

Armed with these refinements, the dominant orthodoxy in evolutionary thinking over the last few decades saw evolution as a 'field of forces'.[6] Selective pressures (the physical metaphor is fitting) act on the genetic variety of a population, producing changes over time according to an optimization of the fitness potential. The adaptationist or neo-Darwinian stance comes from taking this process of natural selection as the main factor in organic evolution. In other words, orthodox evolutionary theory does not deny that there are a number of other factors operating in evolution; it simply plays down their importance and seeks to account for the observed phenomena solely on the basis of optimizing fitness. The Origin of the living and its diversity is foundationally referred to this external referent which is the landscape of ever-improving fitness.

The challenge to this orthodoxy takes what should be by now a familar form. Stated bluntly: the intrinsic characteristics of life itself are much if not the central driving force in evolution, shaping what counts as fit or not. Thus the Other is seen in this light not as Origin but as result. Once again the logic of *différance* in action. For this re-reading of evolution, modern biologists bring into the foreground a number of factors, all pertaining to the richness of organisms as networks (of traits, of genes, of behaviors). It is these self-organizing qualities which constitute the reverse arrow from what seems to be hierarchically dependent, to become constituting and causal. A partial list of these factors is gene interdependence (pleiotropy), developmental constraints, genetic drift, evolutionary stasis, and the various levels for units of selection. As Richard Lewontin said in a recent critique of the classical position: "It is not that these phenomena [i.e., developmental constraints, pleiotropy, etc.] are not mentioned, but they are clearly diversions from the big event, the ascent of Mount Fitness by Sir Ronald Fisher and his faithful Sherpas."[7] Increasingly, evolutionary biologists have become engaged in a movement away from Mount Fitness towards a larger and as yet incompletely formulated new theory.[8] Alfred Russell Wallace was fond of saying that "Nothing in Nature is not useful."[9] The mood of our culture still agrees with, and many scientists use, without even so much as a second thought, arguments that appeal to the parsimony of nature and its optimal designs. The list of tangled issues that we have just discussed, however, indicates otherwise.

All these themes are touched in the text by Kauffman and in the comments addressed to it. Kauffman's approach is very ingenious, for he focuses on a clear case of adaptation, a space of small proteins, and proceeds to complexify it until the point where the reverse, paradoxical effects of internal factors begin to appear. Although his arguments by no means prove a grand unified theory, they are some of the best yet produced in clarity and rigor to explore what a post-Darwinian theory could look like.

Brian Goodwin adds to the themes raised by Kauffman by focusing on another dimension which borrows from another central but much neglected discipline in evolution: development. His main point is that developmental constraints are so important and pervasive that a grammar of the organism is much the same as what one sees in the history of the biosphere. Daniel Brooks contributes an original angle to the question of revising neo-Darwinism by examining the circulation of energy in the transformation of matter, and thus seeing concepts of entropy and 'information' as a common framework for developmental and evolutionary processes. Susan Oyama and John Dupré, on the other hand, continue the debate in a more philosophical tone. Dupré's point is to evaluate Kauffman's arguments and examine whether they are still not relying far too much on some adaptationist framework. His plea is for a multifaceted evolutionary theory with room for various multilevel mechanisms. Oyama addresses the important issue of the nature-nurture opposition in the light of post-Darwinian theories. As she points out, such opposition is always indicative of a severe demarcation line between an outside and an inside. When these two are interlinked — as the logic of the supplement demands — the opposition disappears into another conceptual landscape.

To conclude then, let us reiterate: the crux of the matter is that to explain an observed biological regularity as an optimal fit or optimal correspondence with pre-given dimensions of the environment appears less and less tenable on both logical and empirical grounds. Part of the difficulty in moving beyond the neo-Darwinian framework is to determine what to do after we abandon the idea of natural selection as the main explanation, so that every structure, mechanism, trait or disposition cannot be accounted for by its contribution to survival value. The temptation is to say: "But, then, are things there for no reason at all?" The task of a post-Darwinian evolutionary biology is to change the logical geography of the debate by studying the tangled, *circular relations* of congruence among the items to be explained. This lands us squarely with the same kind of logic as evoked before, as sketched in Fig. 6.

According to traditional wisdom, the environment in which organisms evolve and which they come to know is given, fixed, and unique. Here again we find the idea that organisms are basically parachuted into a pre-given environment. This view undergoes refinement when we allow for changes in the environment, an allowance that was already empirically familar to Darwin. Such a moving environment provides the

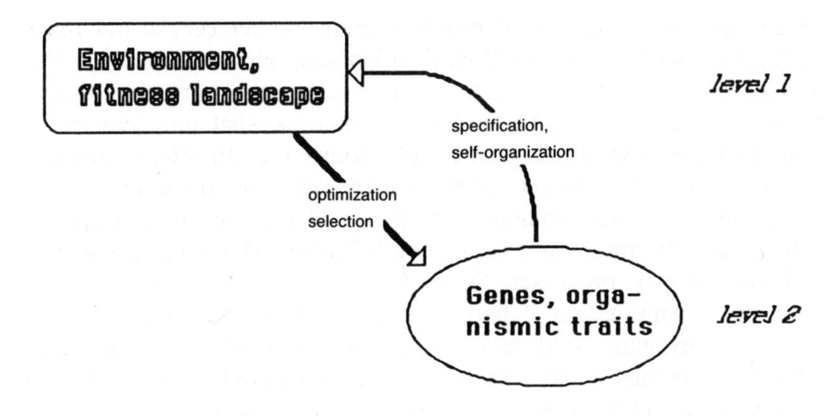


Fig. 6.

selective pressures that form the backbone of neo-Darwinian evolutionary theory. In moving towards a revised view of evolution, however, we introduce a further step: we recast selective pressures as broad constraints to be satisfied. The crucial point here is that we do not retain the notion of an independent, pre-given environment, but let it fade into the background in favor of so-called 'intrinsic factors'. Instead, we emphasize that *the very notion of what an environment is cannot be separated from what organisms are and what they do*. This point has been made quite eloquently by Richard Lewontin:

> . . . the organism and the environment are not actually separately determined. The environment is not a structure imposed on living beings from the outside but is in fact a creation of those beings. The environment is not an autonomous process but a reflection of the biology of the species. Just as there is no organism without an environment, so there is no environment without an organism.[10]

The key point, then, is that the species brings forth and specifies its own domain of problems to be 'solved' by 'satisfying'; this domain does not exist 'out there' in an environment that acts as a landing pad for organisms that somehow drop or parachute into the world. Instead, living beings and their environments stand in relation to each other through *mutual specification* or *co-determination*. Thus what we describe as environmental regularities are not external features that have been internalized, as representationism and adaptationism both assume. Environmental regularities are the result of a conjoint history, a congruence which unfolds from a long history of co-determination. In Lewontin's words, the organism is both the subject and the object of evolution.[11]

The final topic addressed during the Conference was the origin of cognitive content, of the meaning of the world as we perceive it. More specifically, the discussions were really concerned with basic cognitive capacities: perception and action, rather than with so-called higher capacities such as language and memory.

The approach followed here in Varela's text is to view cognitive science as consisting of three successive stages. The three stages correspond to three successive movements; each indicates an important shift in the theoretical framework within cognitive science. The center or core of cognitive science is known generally as *cognitivism*.[12] The central tool and guiding metaphor of cognitivism is the digital computer as a physical device built in such a way that a particular set of its physical changes can be interpreted as computations. A computation is an operation performed or carried out on symbols, i.e. on elements that *represent* what they stand for. Simplifying for the moment, we can say that cognitivism consists in the hypothesis that cognition — human included — is the manipulation of symbols after the fashion of digital computers. In other words, cognition is *mental representation*: the mind is thought to operate by manipulating symbols that represent features of the world, or represent the world as being a certain way. According to this cognitivist hypothesis, the study of cognition *qua* mental representation provides the proper domain of cognitive science, a domain held to be independent of neurobiology at one end, and sociology and anthropology at the other.

Cognitivism has the virtue of being a well defined research program, complete with prestigious institutions, journals, applied technology, and international commercial concerns. We refer to it as the center or core of cognitive science because it dominates research to such an extent that it is often simply taken to *be* cognitive science itself. In the past few years, however, several alternative approaches to cognition have appeared. These approaches diverge from cognitivism along two basic lines of dissent:

(1) A critique of symbol processing as the appropriate vehicle for representations.

(2) A critique of the adequacy of the notion of representation as the Archimedes point for cognitive science.

The first alternative, which we may call 'emergence' is typically referred to as connectionism. This name is derived from the idea that many cognitive tasks (e.g. vision and memory) seem to be handled best by systems made up of many simple components, which, when connected by the appropriate rules, give rise to global behavior corresponding to the desired task. Cognitivist symbolic processing, in contrast, is localized: operations on symbols can be specified using only the physical form of the symbols, not their meaning. Of course, it is this feature of symbols that enables one to build a physical device to manipulate them. The disadvantage is that the loss of any part of the symbols or the rules for their manipulation results in serious malfunction. Connectionist models, on the other hand, generally trade localized, symbolic processing for distributed operations, i.e., ones that extend over an entire network of components, and so result in the emergence of global properties resilient to local malfunction. For connectionists a representation consists of the correspondence between such an emergent global state and properties of the world; it is not a function of particular symbols.

The second alternative is born from a deeper dissatisfaction than the connectionist search for alternatives to symbolic processing. It questions the centrality of the notion that cognition is fundamentally representation. Behind this notion stand two fundamental assumptions. The first is that we inhabit a world with particular properties, such as length, color, movement, sound, etc. The second is that we 'pick up' or 'recover' these properties by internally representing them. These two assumptions amount to a strong, often tacit and unquestioned, commitment to realism or objectivism about the way the world is and how we come to know it.

There are, however, many ways that the world is — indeed even many different worlds of experience — depending on the structure of the being involved and the kinds of distinctions it is able to make. And even if we restrict our attention to human cognition, there are many various ways the world can be taken to be.[13] This non-objectivist conviction is slowly growing in the study of cognition. As yet, however, this alternative orientation does not have a well established name, for it is more of an umbrella that covers a relatively small group of people working in diverse fields. We propose as a name the term *enactive* to emphasize the growing conviction that cognition is not the representation of a pre-given world, but is, rather, the enactment or bringing forth of a world on the basis of history and the variety of effective actions that a being can perform. The enactive approach takes seriously, then, the philosophical critique of the idea that the mind is a mirror of nature, but goes further by addressing this issue from within the heartland of science.[14]

At this point the reader might have noticed that each alternative to the study of cognition enlarges the domain of cognitive science. At each stage, then, we focus on something we had previously taken at face value, such as a symbol or a representation. Finally, at the periphery we become explicitly concerned with the origin of regularities that usually seem stable and fixed, and so amenable to a symbolic or representational treatment. Having reached this point, we can then move in the reverse direction, from periphery to center, by bracketing a concern with the origin of those regularities, to consider them simply at face value.

The text by Christine Skarda complements the discussion by Varela, in another light. The basis of her remarks is the important work by Walter Freeman who has introduced the notion that the dynamical constitution and properties of neuronal networks are in themselves rich enough to question the idea that the brain works by representations in the classical sense. The dynamics of neural ensembles is today a very active pole of research, and Skarda proposes that the dynamics vs. connectionist/symbolic distinction is the most relevant one for the origin of perception.

It is interesting to remark how the received view of evolutionary processes, as described before, can be in fact understood by the representationist idea that there is a correspondence between organism and environment provided by the optimizing constraints of survival and reproduction. Baldly stated, representationism in cognitive science is the precise *homologue* of adaptationism in evolutionary theory, for optimality plays the same central role in each domain. It follows that any evidence that weakens the adaptationist viewpoint *ipso facto* provides difficulties for the representationist approach to cognition.

Now, cognitive scientists have been relentlessly led by the requirements of their research to the study of sub-networks that act on local scales. These networks interact with each other in tangled webs, forming 'societies' of 'agents', to use Minsky's language. It should be clear from our list of current problems that evolutionary theorists have independently reached much the same conclusions. The constraints of survival and reproduction are far too weak to provide an account of how structures develop and change. Accordingly, no *global* optimal fitness scheme apparently suffices to explain evolutionary processes. There are, to be sure, local genetic 'agents' for, say, oxygen consumption or feather growth, which can be measured on some comparative scale where optimality may be sought, but no single scale will do the job for all processes.[15] We are thus back to our familiar logic of the supplement for cognitive phenomena, as illustrated in Fig. 7.

The central issue for both biological instance, evolution and cognitive meaning, can be put in the form of an analogy.[16] John needs a suit. In a fully symbolic and representationist world, he goes to his tailor who measures him and produces a nice suit according to the exact specifications of his measurements. There is, however, another obvious possibility; one which does not demand so much from the environment. John goes to several department stores and chooses a suit that fits well from among the various ones available. Although these do not suit him *exactly*, they are good enough and he chooses the optimal one for fit and taste. Here we have a good selectionist alternative that uses some optimal criteria of fitness. The analogy admits, however, of further refinement. John, like any human being, cannot buy a suit in isolation

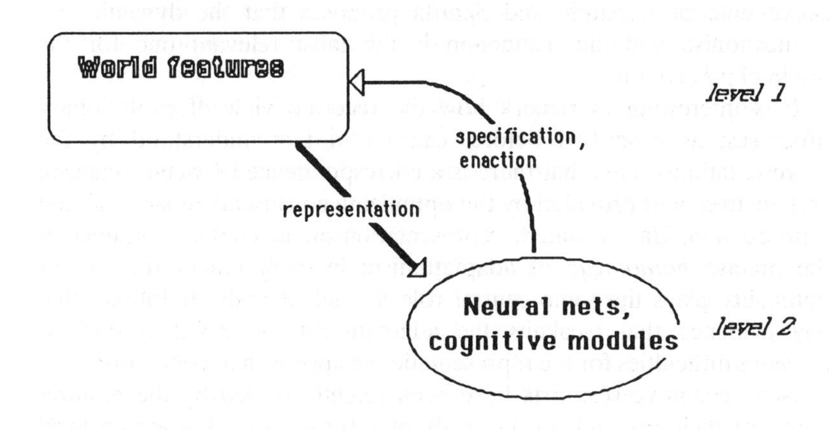


Fig. 7.

from the rest of what goes on in his life. In buying a suit, he considers how his look will affect the response of his boss at work, the response of his girlfriend, and he may also be concerned with political and economic factors. Indeed, the very decision to buy a suit is not given from the outset as a problem, but is constituted by the global situation of his life. His final choice has the form of *satisfying* some very loose constraints (e.g., being well-dressed), but does not have the form of a fit — and even less so of an optimal fit — to any of these constraints.

Thus speak the case studies we have assembled in this volume. They are all permeated with the tendency to think in terms of what we might call an *exogenous* point of reference or fixed point, or some collection of them. These must be impenetrable and unmovable, and order arises from them as a pre-established harmony with such exogenous fixed points, be these God or His terrestrial substitutes: gold, environmental optimization, representation of traits. But as we have repeatedly seen, such grounding of an origin in an exogenous fixed point is prone, nay, infected, with Derridian corrosive. It is never the case that such externalities can be held up except by the supplement of the order they supposedly engender. Whence the backlash to the other extreme: God is dead, the world is but an eternal chaos, without beauty or origin, without meaning. There is nothing but interpretations of interpretations, *ad infinitum*. "There are no facts, only interpretations; our discourse is our 'new infinity'." The quotations are all from Nietzsche.[17] (Nietzscheism is, indeed, more widespread today than one thinks.)

But that's the interesting paradox. Between the tendency for an exogenous conception of order and nihilism there is a complicity, in spite of their apparently contradictory positions. To go from the first to the second we only need a dose of logic of the supplement, to get rid of what seems to be a fixed point exterior to the system. The passage from the second to the first is less obvious and where nihilism errs: to see that there are principles of endogenous arising, crafted interdependencies that are capable of giving rise to *endogenous* fixed points. This is what we have traced here for the cases of social order, money, evolution and perceptual cognition. Their respective seemingly exogenous fixed points once removed, we can see the establishment of a full circle of self-reference giving rise to a non-arbitrary endogenous constitution. For this endogenous fixed point there is order: society, value, species, and objects in the world. But unlike the claim of the exogenous view of origins, such externalizations are not grounded anywhere except in the constituting agents themselves and the peculiar processes that interlink them. They all partake of a similar fundamental generative logic.[18] When this creative circle is broken and flattened out into a one-dimensional opposition, we recover the antithetical tendencies of exogenous order vs. deconstruction: see Fig. 8.

We thus submit that order and its origin may be thought of — however dimly and incompletely — in a way that escapes *both* the temptation of ultimate grounding as exogenous references into an ultimate Reason or Truth, and the temptation of complete absence of any regularity and disconnection from any order at all. The key is in the discovery, for each case and in each domain but sharing in a common logic, of a morphogenetic process capable of self-grounding and self-

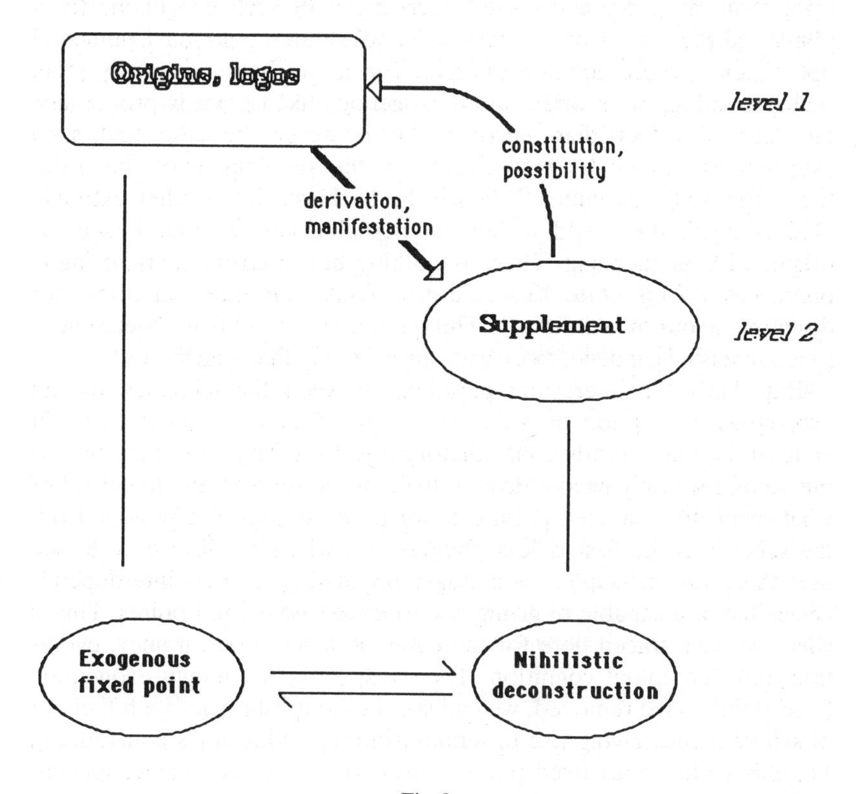


Fig. 8.

distinction. An origin which is neither non-existent or elusive, nor ultimate ground or absolute reference.

*CREA, Ecole Polytechnique, Paris*

## NOTES

[1] H. Maturana and F. Varela, *Autopoiesis and Cognition*, [Boston Studies, vol. 42], D. Reidel, 1980; L. Margulis, *Symbiosis in Cell Evolution*, Freeman, 1982.
[2] F. Varela, *Principles of Biological Autonomy*, North-Holland, New York, 1979.
[3] The ones we have in mind here are *Deceit, Desire, and the Novel*, The Johns Hopkins U. Press, 1965; *Violence and the Sacred*, ibid., 1977; *Des choses cachées depuis la fondation du monde*, Grasset, 1978; *Le Bouc émissaire*, Grasset, 1982.
[4] The term *différance* does not exist in French, and was introduced by Derrida to designate both a difference and deferral.
[5] See in particular: S. J. Gould, 'Darwinism and the expansion of evolutionary theory', *Science* **216** (1982): 380—387; and S. J. Gould and R. Lewontin, 'The Spandrels of San Marco and the Panglossian Paradigm: A critique of the adaptationist programme', *Proceedings of the Royal Society of London* **205** (1979): 581—598. For more general discussion, see Eliot Sober *The Nature of Selection* (Cambridge, Mass.: MIT Press, 1984); M. Ho and P. Saunders, *Beyond Neo-Darwinism* (New York: Academic Press, 1984); and J. Endler, 'The newer synthesis? Some conceptual problems in evolutionary biology,' *Oxford Surveys in Evolutionary Biology* **3** (1986): 224—243. For a recent *defense* of Neo-Darwinism in the face of these various challenges see: M. Hecht and A. Hoffman, 'Why not neo-Darwinism? A critique of paleobiological challenges?' *Oxford Surveys in Evolutionary Biology*, **3** (1986): 1—47. Our discussion in this section also owes much to M. Piatelli-Palmarini, 'Evolution, selection, and cognition,' in E. Quagliariello, G. Bernardi, and A. Ullman (eds.), *From Enzyme Adaptation to Natural Philosophy*, (Amsterdam: Elsevier, 1987), which explores similar themes, though in the context of a defense of cognitivism.
[6] This term is from Eliot Sober, *The Nature of Selection* (*op. cit.*).
[7] R. Lewontin, 'A natural selection: Review of J. M. Smith's *Evolutionary Genetics*,' *Nature* **339** (1989): 107.
[8] An interesting example of this revisionist mood is the critical study of the classic example of industrial melanism in moths as a textbook case of natural selection. According to D. Lambert, C. Millar, and T. Hughes, 'On the classic case of natural selection', *Biology Forum* **79** (1986): 11—49, this example can be transformed into a classic study against neo-Darwinism by considering a substantial amount of ignored extant literature.
[9] H. Clemens, *Alfred R. Wallace: Biologist and Social Reformer* (London: Hutchinson, 1983).
[10] Richard Lewontin, 'The organism as the subject and object of evolution', *Scientia* **118** (1983): 63—82.
[11] *Ibid.*

[12] This designation is justified by John Haugeland, 'The nature and plausibility of cognitivism', reprinted in *Mind Design: Philosophy, Psychology, Artificial Intelligence*, John Haugeland (ed.) (Montgomery, Vt.: Bradford Books, 1981). Sometimes cognitivism is described as the 'symbolic paradigm' or the 'computational approach'. We take these designations as synonyms for our purposes here.
[13] See Nelson Goodman, *Ways of Worldmaking* (Cambridge/Indianapolis: Hackett Publishing Company, 1978).
[14] See Richard Rorty, *Philosophy and the Mirror of Nature* (Princeton University Press, 1979).
[15] For the best thorough and technical discussion of this point see G. Oster and S. Rocklin, 'Optimization models in evolutionary biology', *Lectures in Mathematical Life Sciences*, vol. 11 (Rhode Island: American Mathematical Society, 1979), pp. 21—88. For recent discussion see J. Dupré (ed.), *The Latest on the Best* (Cambridge, Mass.: MIT Press, 1987).
[16] This analogy was first proposed in G. Edelman and W. Gall, 'The antibody problem', *Annual Review of Biochemistry* **38** (1979): 699—766. It is also used by M. Piatelli-Palmarini, 'Evolution, selection, and cognition', in E. Quagliariello, G. Bernardi, and A. Ullman (eds.), *From Enzyme Adaptation to Natural Philosophy*. We use the analogy here with an extension which is not in line with the intention of either of these authors.
[17] Cf. § 109 and 374 *Gay Science*, and II-§ 133 *The Will to Power*.
[18] More on this logic in J.-P. Dupuy, 'Tangled Hierarchies: self-reference in philosophy, anthropology, and critical theory', *Comparative Criticism* **12** (1990): 105—123.

RENÉ GIRARD

# ORIGINS: A VIEW FROM THE LITERATURE

The best I can do within the space allotted to me is to open the discussion of our first topic, the origin of society, along the lines of Andrew McKenna's paper. Most people take for granted that the deconstructive critique of philosophical origins is incompatible with the theory of origin that I defend. Andrew McKenna does not agree, and I generally agree with him.

Derrida has focused on philosophical and theoretical texts that deal with origins, especially the origin of language in Saussure's *Cours de Linguistique générale*, Rousseau's *Essai sur l'origine des langues*, Plato's *Phaedrus*, and also Husserl, Kant, Hegel and many others. Derrida claims that instead of being organized around the principles that they claim for their own, these discussions of origin are patterned on what he calls the *logique du supplément.*

First, let us define this supplementary or supplemental logic, or rather non-logic. From now on, I will say more simply 'the supplement'.

In all the works analyzed by Derrida:

(1) Origins (the rational Logos, nature, living speech, the self) present themselves as pure, self-contained, spontaneous, self-sufficient; they need no help from outside; they are fully present to themselves and in themselves.
(2) Everything else (writing, culture, social life, techniques, adulthood) is added and superimposed from outside. The original origin is said to be too powerful and secure to be adversely affected by the supplement. And yet these secondary additions could be dangerous because they tend to pervert, undermine and supplant the primary origins and they should be avoided, rejected, discarded, expelled.
(3) These additions nevertheless occur and turn out to be indispensable; the primary origins are not as operational and efficient as they should be. The supplement makes up for these unexplained deficiencies and its effectiveness is not explicable either.
(4) Most people find it difficult to acknowledge the relevance of the supplement to the texts that it governs. One should not conclude



*Francisco J. Varela and Jean-Pierre Dupuy (eds), Understanding Origins*, 27–42.

> from this reluctance, however, that they are alien to this logic. Just the opposite; the trickiest part of the supplement is that we ourselves conform to its illogical twists and turns when we are least aware of a possible distortion in our own thinking. The supplement partakes of the *unconscious*, in other words.

Derrida's privileged example is writing; philosophers have always treated *écriture* as a 'dangerous supplement' to verbal speech but, at some point, this *écriture* proves indispensable to the establishment of verbal speech itself even though it is regarded as logically, chronologically and hierarchically second to it.

The paradox is that the secondary origin is crucial to the formulation of the official logic that rejects it and nevertheless, on occasion, calls upon it. Thus, these texts combine two heterogeneous meanings whose cohabitation is as strange as it is necessary.

Now that I have defined the supplement, I will shift to *etiological myths*. They can be defined as the theories of origin of non-scientific culture, and non-philosophical or pre-philosophical cultures. Does the supplement apply to them? Derrida has a few indications that suggest the answer is yes but he has not really focused on them; I have and I can tell you that not only does the supplement apply, but it applies more spectacularly and obviously than it does to philosophical texts.

If you cannot visualize my definition of the supplement which is too abstract, do not worry: myth will make you understand what we are talking about; myths look like the exaggeration and caricature of the logical distortions described by Derrida.

The most common type of myth goes something like this: an invited or uninvited stranger visits the community; deliberately or involuntary, he creates a disturbance and offends or threatens the community; or he may play a nasty prank and is therefore called a trickster . . . .

The community must protect itself or its property and it must get rid of its unwelcome guest. Typically, he or she is driven away and/or killed by the entire community; sometimes the stranger manages to escape, but through supernatural means.

Now comes the paradox. If the community is in a position to be visited by someone, at the beginning of these myths, it must already exist. And yet it must not exist since, after the stranger is expelled, he is perceived as the god or divine ancestor without whom the community would not be what it is, or even would not be at all.

For a specific example, I go to a myth from the island of Tikopia in the Solomon Islands:

> The story . . . is that an *atua*, by name Tikarau came to the land of Tikopia from foreign parts. On his arrival a feast was made, and a huge pile of food [. . .] was [gathered]. Emulation then began between the local [people] and the visitor as to who would be the victor in trials of strength or speed. They had a race. Tikarau slipped and fell. He made a pretence of having injured his leg on one of the rocks and began to limp. Suddenly, however, he made a dash for the place where the provision of the feast lay, and grabbing up the heap, fled to the hills. With the people in close pursuit he made for the crest but [. . .] he slipped and fell once more. The people coming right behind were just able to grab, one a coconut, another a taro, another a breadfruit, and others a yam, before their opponent, gathering himself up, bolted to the edge of the cliff, and being an *atua*, launched himself into the sky and set off for the far lands with his ill-gotten gains. He retained the bulk of the Feast but the people had been able to save for Tikopia the principle foodstuffs, and transmit them to posterity. (Raymond Firth, *Tikopia Ritual and Belief*, Beacon Press, 1968, p. 230.)

Tikarau is a visitor from foreign parts. Even though he is welcome at first, he becomes obnoxious and must be driven away. He is a bad supplement in the exact sense of Derrida; once he is gone, the entire cultural system of the community is implicitly attributed to him.

Tikarau is the principal figure in the totemic system, the only one that transcends the various clans. I have eliminated the names of various places at which he stops on his way out, and these are just as significant for the cultural system as the foodstuffs which he drops, also on the way out. Thus, the bad supplement, the thief, the cheat, becomes the good origin of everything, the founding ancestor but through which process, we do not know; no explanation is given.

The society must pre-date the arrival of this strange hero, since it first welcomes him as a distinguished guest. The totemic foods are already around, in even greater abundance than later, but the totemic system comes into being as a result of the bad supplement turning into a founding ancestor.

The original origin does not seem to prove adequate, since it takes the additional and seemingly negative contribution of the unwanted stranger to make the society be what it really is. The second origin is what the myth is really about and it supplants the primary one.

No reason is given why the dangerous supplement should become the sole origin of a social entity that seemed to need no supplement of origin in the first place.

You see how neatly Derrida's scheme applies to these fellows, the founding ancestors of world mythology. An etiological myth often looks almost like an allegory of the supplement, or a didactic fable, perhaps, devised by eager disciples for the teaching of their no less eager students.

In my reading of etiological myths for the past twenty years, I have always found the same distortions in our logic that Derrida finds in texts of philosophical origin. I have been aware of all this in a vague way, but I have failed to explore these similarities. Philosophy is just too complicated for me. I badly need such people as Andrew McKenna and Cesáreo Bandera, who has also written on our topic and similar topics.

The similarities between mythical and philosophical texts of origin are amazing, but they do not lead me to the same conclusions as Derrida.

I see the supplement not as a reason to despair of Western thinking, but as a potential source of knowledge. My basic attitude toward the possibility of real knowledge is not that of a deconstructor; some of the reasons for the difference are personal, no doubt, but others are not, I think, and they are more interesting: they reflect differences between the cultural domains in which the deconstructors operate, philosophy and the domain which interests me primarily, mythology and ritual.

Deconstruction is the affair of philosophers reacting against the tradition of German idealism and phenomenology in which they have been raised. If philosophical texts obey a (non)-logic contrary to their own stated logic, they conclude that less rationality is available to us than was formerly believed.

Compared to philosophy, mythology sounds like a hopelessly garbled message and the same observation that looks like a curse to philosophy, seems like a blessing in mythology. In the rational context of philosophy, the supplement seems like disorder; in the irrational context of myth, it looks like a potential source of order.

The student of myth generally knows that his texts are full of logical inconsistencies, and the supplement confirms this. What he generally does not know is how consistent these inconsistencies are. The valuable aspect of the supplement, or whatever you want to call it, is to show that the logical inconsistencies are organized in a most consistent pattern.

Just think of this. The supplement is not a vague and all embracing concept, some *idée générale* or cloudy archetype from which we would learn nothing because it could apply to everything. It is a specific and rather complex scheme; and yet it fits not only Polynesian myths but American and African myths, as well as Greek myths, Indo-European myths. It fits the crudest as well as the most sophisticated myths.

And it also applies, of course, to countless philosophical and other texts. The analysis of the supplement is the most original contribution of deconstruction and it cuts across an amazing range of cultural phenomena. I think this ambition is justified, but the deconstructors are never suspected of the cardinal sin of our world, 'reductionism'. With some of our people, the most unavoidable process of abstraction, the very type of generalization that makes you able to walk into the street without being run over by a car, is already tainted with the impurity of reductionism.

You are a 'reductionist' above all if you pursue the type of goal that any researcher outside the Humanities takes for granted that he should pursue.

The deconstructor escapes this reproach because his emphasis is essentially critical in a critical age. He tells us that we cannot really think, and most people do not find this threatening at all, but reassuring.

When the student of myth discovers the supplementary structure of mythology, he can regard it only as a starting point for research, not as an achievement that would be sufficient unto itself. If he announced to the world that mythology has been found illogical and is therefore in the process of being deconstructed, people would ask: what else is new?

Consistent inconsistencies are the very type of clues that students of myth should be looking for. The supplement is precisely that. Once we have perceived that there is a mythical process of sorts and that it consists of what Derrida calls the supplement, we are more able to resist the two great symmetrical deviations that have vitiated the interpretation of mythology for centuries.

The first one is excess rationalism; the second one is excess irrationalism.

The first approach is marked by an excessive reliance on the principle of identity. It is the no-nonsense approach; it unconsciously does away with the logical distortions and it forces myth back into the mould of our own explicit reason. (An example would be Lévi-Strauss.)

The second approach is the opposite and it is the same. The logical distortions are acknowledged this time, but in haphazard and fragmented fashion; the system that they form is not recognized and they are not clearly distinguished from the extreme thematic diversity of mythology. Myth is seen as pure caprice and all attempts at a rigorous treatment are abandoned.

The irrationalists always hurry to produce some universal scheme that is still the child of rationalism in the sense that it contradicts its conception of reason only in the manner that this reason expects to be contradicted. An example is Jung, who sees myth in terms of a badly needed spiritual complement to the thinking that prevails in our society, thereby reaffirming the non-mythical character of whatever calls itself science; and Jung reinforces the scientistic prejudices of our society.

A second example is the current ideology, according to which each myth is uniquely singular and uniquely expresses the uniqueness of each culture. Metaphysically, this ethnological subjectivism is a transposition of romantic subjectivism to the cultural domain. In the name of anti-dogmaticism, it dogmatically denies the possibility of mythical invariants and a theoretician of myth who embraces it is really committing theoretical suicide.

Both the mythologists who surrender directly to the scientistic illusion, or those who surrender indirectly through the collective subjectivism of uniqueness, end up reinforcing the dominant prejudices of our society.

When we do not understand a system of representations, it may be useful to look for a type of experience that is relatively intelligible to us and that could be productive of these same representations or of similar ones.

If you look at etiological myths comparatively, you will discover that, in spite of their fantastic features, and in part because of these, they strikingly resemble the type of representations we associate with mobs on the rampage. Try to put yourself into the frame of mind of a small mob on the rampage and you will find that you are looking for a victim and once you are finished with that victim, you will recollect your deed in a manner that can only be described as mythical or quasi-mythical.

I must cut a long story short and limit myself to a few indications. Many people here do not know what I am talking about, but quite a few do.

The first major clue to the dominance of the mob spirit over myths, at least in a first phase, is that so many of them represent collective violence against a single individual and that this violence is invariably justified by the threat that this single individual would constitute for the collectivity, a most unlikely threat to say the least. This recurrent representation suggests that the violent mob is the subject of the myth and that it enthrones its own collective delusion as truth.

These first two clues (one direct, one indirect) are confirmed by several others. One is the type of threat that the victim represents. It can be quickly identified as the type of accusations that accompany the more or less arbitrary selection of their victims by mobs in a state of panic, or rage, or a combination of both.

In many myths it is the evil eye; by simply looking at the members of the community, the hero makes them drop dead, or he slowly poisons them — Tikarau is accused of theft; numerous mythical heroes are accused of spreading a plague. Typical, too, are sexual offenses and crimes against the family, such as parricide, matricide, all types of incest, bestiality, etc. It can also be a combination of these crimes, as in the case of Oedipus.

If you look at etiological myths carefully, you will see that these accusations typical of mobs are followed by actions also typical of mobs. The accused may be collectively drowned or forced to hurl himself into a fire, or he may be stifled to death by the ever-present mob pressing against him. He may also be trampled to death either by men or by mobs of animals that may well substitute for the human mob; the hero himself or herself may be one of these animals.

In our myth, Tikarau is chased up the hill by the entire community. There is no indication of his being killed, but he flies into the sky, and any observer familiar with capital punishment (so-called) in non-judicial societies knows what a popular form of execution it is in these societies to have the condemned man hurl himself into the void from the top of a cliff — if an appropriate cliff is available, of course.

The scene described in our myth is a possible example; we do not have to decide if the practice must be regarded as already customary in this particular instance or if it is a first, an initial scene, an invention from which ritual custom may or may not originate.

The famous Tarpeian rock is a ritual custom which turns into a judicial mode of execution, the original of which must lie in a Tikarau-like scene which remains unidentified.

We have typical mob accusations plus typical mob behavior. But we have more. Once you are willing to entertain the possibility that mythical heroes might be real victims, you will quickly discover that many of these victims must be not quite random, but only partially so in the sense that they often exhibit features suggestive of mob behavior all over the world. Many of them are afflicted with a limp, such as Hephaistos and Oedipus once again, or they pretend to limp, which is what Tikarau is accused of doing, even though the second time, he seems to really limp.

Many heroes are portrayed as people with a personal trait in society that makes them appear uncanny or simply objectionable; some stammer, others have bad breath, or their entire body exhales a foul odor; others still are hunchback, maimed or blind, or they are accused of faking blindness, etc.

Take all these clues together and you have all possible features of mob prejudice; if a composite image were drawn, the mythical hero would remind us of the village outcast, the abnormal individual who finds himself the butt of collective persecution in a backward community. The only aspects of social discrimination poorly represented in mythology are the ones too specific of our culture, or of all culture, to be discernible in a transcultural context — cultural prejudice as such, religious and ethnic hostility. We can surmise, however, that they are not really missing from the enormously frequent designation of the victim as *a stranger*, and our myth, of course, is an example of that.

If you have followed my observations, you get my general drift. Statistically, there are too many mythical representations characteristic of mobs on the rampage for us to avoid the inference that myths must originate in mobs on the rampage.

If myths do indeed originate in mob delusion it means that some mythical representations are just as false as we always thought they were, but it also means that other representations must be true, in the sense that they must correspond to real information.

This is what most people cannot accept in my interpretation of mythology and yet it is a matter of common sense. The representations that could be true are those that describe a stranger showing up, one fine day, in Tikopia and being well received at first. But then, the "contests of speed and strength" suggest that a spirit of rivalry and conflict arose; the stranger was finally chased up a cliff until he could not flee any further, and then, seized by panic or pushed by the crowd, he went over the top.

One finds scenes of this type all over the world, with too many different types of violent death and too many different concrete details in many cases to support either the idea of some purely symbolic murder or Freud's idea of a single real murder. The only hypothesis that makes sense is that of a different mob on the rampage behind every single myth.

But what about the representations of the myth that are probably or certainly false, the idea that Tikarau stole the totemic food, or even the idea that he stole anything, the idea that he flew into the sky, the idea that he is the origin of the cultural system of Tikopia?

These ideas are unbelievable of course, but the type of unbelievability that they embody, in the specific context of this story, obviously reinforces rather than diminishes the probability that the myth originates in a mob on the rampage: they reflect exactly what we can expect from the mob on the rampage.

You will agree with me, I presume, that Tikarau's theft and his flying into the sky do not pose a serious problem from my perspective. The mysterious aspect is the delinquent's role as a good origin of Tikopia's cultural system, in other words his supplementary metamorphosis. If simple common sense were sufficient to ascribe this idea to a mob effect, we would all understand what the true origin of myth really is.

This is our only difficulty and it is far from insurmountable. The first thing to perceive is that, even though the delusion is more spectacular in myth, it is of the same logical type, the same supplemental type as the historically observable delusions of mobs on the rampage.

We all realize that intense forms of racial, religious or cultural discrimination will turn the members of the targeted groups into bad supplements if they show up among the people who discriminate against them, but this discrimination often plays such an enormous role in the internal cohesion of the discriminators that the rejected outsider is really at the very center of the community that rejects him; he is the pillar of that community insofar as everyone and everything ultimately revolve around him.

Prejudice is a quasi-mythical text never really organized as it is supposed to be. It is too fascinated by its enemies; its own center of gravity is always shifting from the pseudo-narcissistic self-contemplation of the prejudiced group to the dangerous intruder; the more pronounced this shift, the more the bad supplement tends to turn into a good origin. The representations of persecutors truly deluded by their own persecution always tend to organize themselves according to the logic of the supplement.

In order to understand mythology, we have to assume that, instead of remaining covert and requiring a great deal of complex analysis to be uncovered, this tendency is so powerful, at the myth-making level, that it triumphs completely and the mob shifts from distrust and hostility *vis-à-vis* its bad supplement to an attitude of reverence and gratitude. The same victim is now regarded as the origin and substance of what the group is or should be.

In all mobs pacified by their own violence this tendency to gather *around* their victims as well as *against* is observable in some measure. The singularity of myth, especially those myths that seem most illogical to us is that the *against* and the *around* are present side by side explicitly, in a form so extreme that Tikarau is simultaneously responsible for all the disorder and strife that accompanied his violent murder and for the return to peace and order that followed his disappearance behind the cliff.

That is the reason why a primitive god or founding ancestor is a fearsome enemy as well as a powerful friend, a rallying point and yet someone to stay away from, an agent of punishment and retribution as well as of earthly rewards.

Mobs must reach the myth-making stage when their delusion becomes so extreme that their recollection of what they did explicitly reorganizes itself around the formerly expelled outsider as its new center.

If the deluded persecutors believe in the destructive power of their victim intensely enough to be reconciled by its death, they will also believe in the reconciling and reconstructive power of that victim because the tendency to make him the sole agent of the entire episode will apply to the peace that follows, as well as to the turmoil that precedes his death.

The death of a victim in the circumstances of our myth, must have had a sufficient impact on the group chastised by its own violence to trigger not an absolute beginning, assuredly, but a partially new beginning, a rejuvination of the cultural order and a period of social discipline favorable to the creative imitation (ritual), counter-imitation (prohibition), and transfigured recollection (myth) of the model and counter-model that the mob phenomenon has restrospectively become.

An ancestor bad enough to steal the totemic food and good enough to give us our totemic system cannot be an ordinary human being. We can see here why a mythical hero has to be transcendental, supernatural. His power for both evil and good appears such that it requires a superhuman being who in the present case is called an *atua*. The early Tikopians treated him a little roughly no doubt, but he certainly deserved it, and their descendants revere him because they owe him a great deal.

The transformation of the bad supplement into a good origin must correspond to a real social experience, it must be a true representation, to the extent that it must correspond to an improved operation of a renovated cultural system in the group whose origin the myth is supposed to recount. In most etiological myths, there are no indications that an absolute origin is meant.

If you are willing to entertain the hypothesis of the violent mob as the generative engine behind mythology, all mythical representations make sense as well as their supplementary arrangement and they make complete sense as a necessary part of the type of recollection that an extremely deluded mob will have of its own actions.

I believe that etiological myths contain a great deal of real information about certain modifications that the cultural systems undergo and, indirectly, about how all cultural systems originate and transform themselves, but I will not go into that.

The role of mob phenomena does not preclude the possibility of objective factors behind the social crisis often represented in the myth, as well as of borrowings, influences, deliberate tampering, etc. Even if our myth is a mere copy of another myth, and if this model itself is a copy, ultimately there must be a real victim behind some more original model because only a real victim can account for the nature and the supplementary arrangement of the representations that we have in these myths.

I find it ironic that this hypothesis is often condemned for its naive trust in mythical representation but I understand very well that it must inevitably happen.

I say that many mythical representations must be trustworthy in principle, if not in each actual instance, in the sense that they directly reflect the social process in which myths originate. This is so startling that it is dismissed most of the time with not even a glance at the mythical data. My views are in total contrast with the entire modern tradition of mythological studies, which regards myths as purely and necessarily fictional; this dogma is strongly reinforced by the current attitudes toward representation. At a time when it is fashionable to proclaim that the social sciences are pure fiction and even the exact sciences, to say that myths contain a lot of actually true representations sounds like a surrealistic joke.

It is very easy to show that it is no joke by showing that the intellectual operation that this reading of myth represents is the exact duplicate of our most daring yet universally trusted interpretation of mob phenomena productive of a quasi-mythical text in our own Western history.

An official account of a medieval or Renaissance witch trial, is quasi-mythical in the sense that everybody, including the accused, believes that the witch flies into the sky and does many of the same bad things and perhaps some of the good things that founding ancestors did.

Which is the most powerful critique of this account? Is it the one that would deny all informational value to all the representations and treat this text as if it were a libretto of Gilbert and Sullivan, or is it the one that acknowledges the presence of a real woman behind this text, a woman who was really burned at the stake?

In the case of full-fledged myths, as well as in the case of these quasi-myths, we must postulate a certain amount of accurate information in order to achieve a critique of the false representations much more powerful and pertinent than their blanket dismissal as pure fiction.

Four centuries have elapsed since mankind became able to overturn the system of representation of medieval witchcraft and other forms of magical persecution in historical texts, and now the time may be at hand when the same is going to happen with mythology and ritual.

Few researchers, however, are willing to entertain the possibility of a generative and structural hypothesis of myth. The anthropologists are too positivistic, too narrowly empirical, too exclusively focused on 'their own field'. They do not perceive the possibility of reaching, through the comparative method, a level that would be generative.

With the deconstructors, the problem is the very reverse and the final result is the same; they are not empirical enough; they are not interested enough in the data to perceive *how similar myths are to those representations behind which mob effects have become immediately and automatically identifiable.*

One last clue: myths of origin are called etiological; the word comes from the Greek *aitios, aitia.* The first meaning of it is not cause, origin, which comes only in second place in the dictionaries, but culprit, guilty, accused, delinquent.

There seems to be a stage in human thought, not necessarily a more primitive one, during which the only possible cause is a culprit. In many Australian societies, natural death was not supposed to exist. When someone died, there had to be a culprit, a bad supplement from outside who was always selected in one of the neighboring tribes. He was the *aitios* with the dual meaning of culprit and of course, always a little demonic and a little divine.

Either this collective victimage hypothesis accounts for the data or it does not. If it does not, it does not have the slightest interest. I find it distressing that many people condemn it or even sometimes applaud it with no reference to the data, as if its merit or lack of merit depended on some intrinsic virtue, as if it were a gratuitous story of origin, in other words.

This is particularly distressing, of course, in the case of the deconstructors themselves who are the best placed to see that the victimage hypothesis of mythical origin is rooted not in a naive trust in the mythical representations, but in the observation of the same logical distortions that they themselves spot in philosophical texts — but a stronger and clearer version of these, and one, therefore, that should be easy to recognize.

If the supplemental scheme is relevant to myths all over the world, it cannot be characteristic of Western metaphysics only or even primarily. Not that its application to Western metaphysics is false. What I question is its limitation to Western thinking. The supplement must reach something fundamental in the human thinking process as a whole.

I said before and now I repeat that myths look like a caricature of the supplementary distortions of philosophical origins, but that cannot be entirely correct. The word caricature implies that philosophy would be the original and myth would be the copy. The word distortion cannot be right either. It seems to imply that the logic of identity was first and was then distorted by mythology. If there is a relationship of dependence, it can only go the other way around; philosophy could be dependent on myth but not the reverse. The logic of the supplement must be mythical first and philosophical second. Philosophical origins could be a weakened version of what we have in myth.

Mythology seems like a caricature of the supplement because it reflects this general human inaptitude to self-centeredness, this failure of individual and collective narcissism, and the resulting fear of and fascination with otherness in a most brutal, fundamental and revealing form. It also suggests a positive side of this weakness, its role in the creation of culture, in the development of human thought.

The role of human relations in all of human thought may be greater than has been realized until now. Philosophy may reflect forms of conflict and intellectual competition quite different, of course, from mob phenomena, but not entirely unrelated to them. If I had to put a general label onto the relational background of philosophy, I would call it polemical.

Derrida, after many others, shows this polemical background in the case of Plato. The polemics of philosophy are continuous, perhaps, with the violence of mythology, and this continuity may be one of the meanings of Heraclites' famous saying about *Polemos* being father and king of everything.

We all know in academic life that human thought is fundamentally polemical at all times and, if it cannot find any enemy outside of itself, it will take itself as a victim and self-destruct, as it often does in our time.

I think that many effects of structural de-centering that can be observed in intellectual systems are inseparable from this polemical dimension, and this dimension is not only a rejection but also a fascination of philosophy for its other. All human structures try to have their own center in themselves, to be self-centered, but they cannot; they are de-centered. This de-centering is another theme of our time; Derrida wrote an essay on the de-centered structure. A de-centered structure is an unstable structure that first tries to resist what it regards as an outside aggression and will even react violently against it, but it can shift around and re-organize itself around what it still regards as an inimical and hostile force.

The so-called supplement represents in a static structure, a synchronic configuration, a process of crisis and restructuration that remains too incomplete and too insufficiently temporalized to appear as two separate structures, and we have not yet learned to read the various modalities of the supplement as a trace of the passage from one to the other. I leave it to Andrew McKenna to tell me how much Derrida there is in that last sentence of mine.

A text will organize itself as a supplementary machine if the writing of it reflects a hesitation between two centers, the author's official center of gravity and the object of his fascination, of what he deliberately rejects.

I will give you a highly simplified, even simplistic example, but it will make you understand what I am talking about: take American intellectuals during the Cold War. Marxism is the bad supplement, the useless addition, but if you look more closely you will see that even in the fifties, American political science was already influenced by it; it tends to see intellectual phenomena in economic terms and to explain all non-economic value as a reflection of economic realities.

Take Russian intellectuals during the same period or a little later. The free market economy is the official enemy and everything American is regarded as a hostile intrusion, a bad supplement. If you have a closer look at the way these people think, you will discover that they, too, are strongly influenced by what they reject. Each system's center of gravity really lies on the other side of the fence. That may be the reason why, today, we have more Marxist researchers in America than in Russia, and more free enterprise devotees in Russia than in America.

I am giving a slightly caricatural example because I am in a hurry and the caricature enables one to understand very quickly a schematic version of what I mean. I truly believe that even the most profound and original forms of thinking embody supplemental aspects because they all have an absolute other that they reject, and this other tends to reappear and play a central role in their thought. Whatever we try to suppress and repress most ardently, our scapegoat, keeps coming back in a manner analogous to but not identical with the so-called return of the repressed in Freud.

In order to show this on a powerful thinker, it takes a tremendous amount of subtle analysis. This is what Derrida has done, I believe, on one of the philosophers in which he traces the supplement most convincingly, Plato. And he reveals a substructure of conflict between two schools of philosophy, that of Socrates and that of the Sophists, which is the conflictual substratum around which the whole supplementary structure revolves. The Sophists are the bad intruders, the useless supplement, but ultimately there are great similarities between their logos and the Platonic logos: Plato, for all his genius, is not a solitary genius; he is fascinated by his philosophical enemies and determined by them at least in part.

Derrida centers his analysis on a family of words that he himself presents as formidably significant, the family of the word *pharmakos* which is the ritual human victim of the Greeks, the human equivalent of the biblical scapegoat. Derrida shows that the word *pharmakon* is used in such a way as to arbitrarily favor Socrates at the expense of the Sophists in his polemical battles with them. The word is an extraordinary link between philosophical polemics and the mob effect of mythology and ritual.

The structure of human thought when it cannot express itself as process and tries to do it as static structure, will always be supplementary because its polemical dynamism will always feed on what it rejects; it will always be obsessed with the other against which it reacts and which it imitates. Non-mythical forms do not surrender to this other as completely as in the case of myth but nevertheless turn it into some kind of foundation scapegoat, too, an antagonist that they cannot do without and that becomes an essential part of the structure in the foundation of which they are buried, and whose presence will reassert itself when it is least expected. They are the mediators of internal mediation. They always play a fundamental role in our own thinking. The dual meaning of the word 'discrimination' in English reminds us of this fundamental role that violent exclusion plays in the very process of human thinking.

*Stanford University*

# PART I

# VIOLENCE: THE ORIGIN OF SOCIAL ORDER

ANDREW J. McKENNA

# SUPPLEMENT TO APOCALYPSE: GIRARD AND DERRIDA

We might begin with Jean-Jacques Rousseau in the eighteenth century. For that is where our problematic quest for origins begins, with the discrediting of Biblical authority, of Scriptural orthodoxy, in terms of which, origins, being divine, are by definition no problem. This in part is the point of Rousseau's insistence, towards the beginning of his *Discours sur l'origine de l'inégalité parmi les hommes*, that "We must begin by discarding all the facts" (III: 132). The context of this remark clearly suggests that he is sidestepping the authority of Biblical narrative, *la lecture des livres sacrés*, which officially at least represents the facts for his culture. This in turn appears to be no problem for our resolutely profane, scientific culture, though part of my argument will concern a relation to the sacred informing and deforming our own quest for origins.

For Jacques Derrida, Rousseau's works provide the occasion for a radical critique of the very notion of origin as well as of our commonplace — and philosophical — notion of representation as the representation of being, of being present, or of being as a presence in which representation originates. What is at issue in Derrida's critique is the capacity of language to refer to anything outside itself, to represent a 'reality' which is not always already a representation, indeed a *supplement* to writing. For writing is conceived in our culture, the most writerly, the most bookish culture of all, as a mere technical supplement to language, an arbitrary representation of representation which veils or distorts our access to reality, the thing itself; whereas for Derrida writing represents a 'supplement of origin', which replaces an origin, a presence which never took place. What is at issue in this regard is the possibility of knowledge, of science itself, along with the notions of truth and value with which every scientific quest is invested. As he states towards the very beginning of *Of Grammatology*, "The idea of science and the idea of writing — therefore also of the science of writing [= grammatology] — is meaningful for us only in terms of an origin and within a world to which a certain concept of the sign (later I shall call it *the* concept of the sign) have *already* been assigned" (1976, 4).



*Francisco J. Varela and Jean-Pierre Dupuy (eds), Understanding Origins*, 45–76.

In *Of Grammatology*, Rousseau is the occasion to challenge this concept, to reveal its internal contradictions. It is these very contradictions that interest us in turn in the exploration of other possibilities in the human sciences, possibilities for a science of man which is rooted in his relations not primarily with signs but with others, as well as in man's relations with what he deems to be wholly other, the sacred by name. I do not mean anything theological by the sacred, which in the anthropological perspective of this essay is but the name which men give to their own violence. I am drawing of course from the 'fundamental anthropology' as advanced by René Girard, principally in *Violence and the Sacred* and in *Des Choses cachées depuis la fondation du monde.* For the contradictions in Rousseau's texts are especially of interest for what they both reveal and disguise, mark and erase about the violence of human origins in its relation to desire. I shall argue that the contradictory logic of the sacred, as it originates in a desire of untraceable origin, conforms to the contradictory logic of the supplement which Derrida uncovers in Rousseau: the supplement of an origin which never took place, the supplement which takes the place of the origin.

Rousseau's desire for origins, to recover them or retrieve them in nature or with a mother or in a self harmonious with a mother/nature, is legible throughout his works. His *Confessions* announce the definitive Revelation of an absolutely original self which he declares would serve as the "primary basis of comparison for the study of men, which has yet to begin" (Preface). By the third paragraph, he summons the Last Judgement to testify to the integrity of his portrait and of its model, while the penultimate paragraph of this work holds its challengers liable to death by smothering. In a subsequent section, we will examine these and similar claims as they testify to a desire which bears all the marks of the sacred as it originates in violence. For the present, Rousseau serves as but the occasion, the pretext, for us to relate Derrida's critique of origins to Girard's originary hypothesis.

> Every enhancement of life enhances man's power of communication, as well as his power of understanding. Empathy with the souls of others is originally nothing more but a physiological susceptibility to suggestions: 'sympathy' or what is called altruism, is merely a product of that phychomotor rapport which is spirituality (*induction psycho-motrice*, Charles Féré thinks). One never communicates thoughts: one communicates movements, mimic signs, which we then trace back to thoughts. (Nietzsche)

What first takes place, according to Girard, is violence, and what takes its place is the sacred. For that we need a victim. The victim is the issue of violence and the origin of the sacred; it comes after nature and before culture, which originates in the sacred, that is in the deference paid to the victim. The victim in this conception both serves to bridge the gap between nature and culture and to mark their definitive rupture. And the difference between nature and culture issues from the primordial difference between man and the sacred, which is inhabited from the beginning by what Derrida neologizes anarcheologically as *différance*.

In the beginning, before the beginning of man, was violence, which has no origin. It is aboriginal, it is predatory. It is especially prevalent in advanced species, whose dominance patterns are fragile and susceptible to breakdown (Hamerton-Kelly, 123—5). It is the mimetic violence of all against all, which has no other origin than another's violence, which it imitates. In the beginning is imitation, not an origin. In the beginning are violent doubles, multiples. It is when the violence of all against all becomes the violence of all against one that a victim is produced, which is likely to happen when a difference, a weakness, marks out a single member of the mêlée for destruction. So in a sense we can say with Derrida that in the beginning was the mark, the trace of a violence which has no origin except in another's violence, a trace of non origin or an arche-trace. We will say just that when we have traced the cultural destiny of the victim. For the victim is the trace of a violence which has no origin except in 'itself', that is in another's violence.

In the beginning, consequent to natural violence, was the victim. Around the victim, is poised a circle of violent predators, each bent on appropriation of its remains. There is nothing more natural than to appropriate the victim for oneself and carry it off. This impulse to violent appropriation is natural to all; it is what produced the victim in the first place; all participate in it — except, in the end, the victim, who represents a difference prior to any identity. But there is nothing more natural either to hesitate before appropriating the victim, lest one fall victim in turn to the unanimous violence which produced it in the first place. No thinking or reflection is required here, only reflex action. This is a structural necessity, as Girard observes in conversation with Walter Burket: "As I imitate my neighbors, I reach for the object he is already reaching for, and we prevent each other from appropriating this object" (Hamerton-Kelly, 123).

It is here that Girard hypothesizes a first moment of "non instinctual attention" (1978c, 109) focusing on the victim; it issues from the 'maximal contrast' between the unanimous *mêlée* which produced the victim and the calm which succeeds its destruction: "As the victim is the victim of every one, the gaze of all the members of the community are in that instant fixed on it. Consequently, beyond the purely instinctual object, the alimentary or sexual object, or the dominant mate, there is the corpse of the victim and it is this corpse which constitutes the first object for this new kind of attention" (1978c, 109). Whether we locate this attention in the spectacular frustration of instinct or in the clash of opposing instincts, the result is the same.

In *The Origin of Language* (Gans, 1981), Eric Gans has refined Girard's hypothesis in order to suggest how this moment coincides with the emergence of the first sign. The natural act of appropriation is a reflex on the part of all; it thereby becomes a gesture, indeed a sign, as it designates the victim as desirable — and as forbidden. For all imitate this movement towards the victim, as all desire the victim, if only out of reflex imitation of movement towards the victim, prior to anything like desire. It is the confluence of these movements which assures the victim's inaccessibility, which indeed makes the victim an object of desire in the first place — or rather in the second place, every desire being a second to another. The object which all desire is perforce the one which none dare appropriate for him- or herself. To designate the victim is to designate it as desirable and as taboo in one and the same movement or moment. The sacred in all its foundational ambivalence — as infinitely desirable and as infinitely accursed, dangerous to approach — is born in and as this same movement or moment of desire.

This moment is contradictory but not incoherent; rather it is 'contradictorily coherent' as Derrida writes of the structural center in 'Structure, Sign and Play in the Discourse of the Human Sciences': "And as always, coherence in contradiction expresses the force of a desire" (1978, 279). Desire is central to Girard's hypothesis, but only as a force which is concentric and eccentric at once.

The victim is holy, sacralized by its deferred possession which alone accords peace to the group. The victim is not desired for its own sake; it has no value which does not emanate from the desires it animates and magnetizes. And its value is contradictory, antithetical — whence the "antithetical sense of primary words" which so fascinated Freud (Vol. XI, 153—162) and which etymo-ethnological research on a word like 'sacer' has born out (Benveniste, 179—207). Sacred words differ from all others in the fact that they differ from themselves from within themselves. They differ, that is, in the mode of *différance*, as Derrida describes it in *Positions*:

> *First, différance* refers to the (active *and* passive) movement that consists in deferring by means of delay, delegation, reprieve, referral, detour, postponement, reserving. In this sense *différance* is not preceded by the originary and indivisible unity of a present possibility that I could reserve, like an expenditure that I would put off calculatedly or for reasons of economy. . . . *Second*, the movement of *différance*, as that which produces different things, that which differentiates, is the common root of all the oppositional concepts that mark our language, such as, to take only a few examples, sensible/intelligible, intuition/signification, nature/culture, etc. (1981, 8–9).

Girard says no less when he attributes to the delay, the postponement, the deferral affecting the victim the root of all subsequent differences:

> Thanks to the victim, such as it appears to come out of the community and as the community appears to come out of it, there can exist for the first time something like an inside and an outside, a before and an after, a community and the sacred. We have already seen that the victim appears at once as bad and good, as peaceful and violent, as life which brings death and as death which assures life. There is no meaning which does not start with it and which at the same time is not transcended by it (1978c, 112).

The verb form in which the victim 'appears' in French is *se présente*. It might be cited as an instance of how " 'everyday language'," according to Derrida, "is not innocent or neutral. It is the language of Western metaphysics", *viz*. the language of being as presence (1981, 19). However the presence or appearing present of the victim in Girard's text is entirely dependent upon abstinence, deferral. It is not ontological presence, still less onto-theo-logical presence that is posited here, but its antithesis: a hallowed illusion, a *quid pro quo*, by which abstinence constitutes presence. Girard's hypothesis in this regard is not one in which the concept of the sign, as sign of a presence, is 'already *assigned*'; on the contrary, he writes, "For the philosophical mentality which still dominates the methodologies of the human sciences, the very notion of hypothesis is inconceivable. Everything remains subject to the ideal of immediate mastery, of direct contact with the *données* which constitute an aspect of what in our day is called a 'metaphysics of presence' " (1978c, 459). Girard does not exploit the ambiguity of language as Derrida does, and this formal difference has substantial implications which remain to be examined. Suffice it for the present to observe that where Derrida's critique of the sign is transposable or translatable in ordinary language, it beats a path to Girard's hypothesis.

The relation of the victim to the community, as just described by Girard, is irreducibly — Derrida would say 'undecidably' (1981, 42–43) — active and passive: the community comes out of the victim no less than the victim comes out of the community. It is transcendent in that it is the matrix of differences and of values (life/death, good/bad, true/false as depending upon being inside or outside the community) which are in no wise antecedent to the power of attraction and repulsion emanating from the victim. "The substitute does not substitute itself for anything which has somehow existed before it" (Derrida, 1978, 411). This power corresponds to Derrida's notion of "the generative movement [of *différance*] in the play of differences":

The latter are neither fallen from the sky nor inscribed once and for all in a closed system, a static structure that a synchronic and taxonomic operation could exhaust. Differences are the effects of transformations, and from this vantage the theme of *différance* is incompatible with the static, synchronic, taxonomic ahistoric motifs in the concept of *structure*. But it goes without saying that this motif is not the only one that defines structure, and that the production of differences, *différance*, is not astructural: it produces systematic and regulated transformations which are able, at a certain point, to leave room for a structural science. The concept of *différance* even develops the most legitimate principled exigencies of a 'structuralism'. (1981, 27–28)

*Différance* is not genetic — except in the sense that it generates all structures and governs their transformations; and it is not structural — except that no principled structuralism is possible without it. It corresponds to Girard's demand for a science of man which is both genetic and structural (1985, 159; Hamerton-Kelly, 108) without privileging either the differences composing a structure nor a chronology imposed by a unique, selfsame origin, a hieratic or hierophanic center which would escape the play of differences. What Derrida objects to in the concept of a *structure centrée* is the concept of an *immobilité fondatrice*, which, bearing the name of God, man, nature, substance, consciousness, etc., would be 'beyond the reach of play' (Derrida, 1978, 279–80). At the center for Girard is *une méconnaissance proprement fondatrice* which is what "opens the way for Difference as such" (1978c, 52).

The origin is double, a relation of mimetic doubles to an object of desire which is antithetically double in its significance, in its value. "Everything begins with structure, configuration, relationship", as Derrida writes paraphrasing Lévi-Straussian structuralism in 'Structure, Sign and Play in the Discourse of the Human Sciences' (1978, 286). So too for Girard, with the decisive difference that the victim makes: all structure emanates from the originary relation to the victim.

The victim is decisive in its very undecidability; it is a homicide and deicide at once – from which the notions of 'homo' and 'deus' subsequently derive their meaning. The gods created man; so says mythology. Man created the gods; so say our human sciences. We have to conceive of the origin of the gods and of man in the same moment, with the gods being constructed retrospectively, belatedly – Derrida would say *après coup, nachträglich* (1978, 210) – out of the deference paid to the sacralized victim, which is but the deference desire pays perforce to an object desired by others. The victim is not 'transcendental signified', which Derrida describes as "*a concept signified in and of itself*, a concept simply present for thought, independent of a relationship to language, that is of a relationship to a system of signifiers" and which "in and of itself, in its essence, would refer to no signifier, would exceed the chain of signs and would no longer itself function as a signifier" (1981, 19–20). Rather Girard describes the victim as 'transcendental signifier' whose signified is "all the actual and potential meaning which the community confers on the victim and, by its mediation, on all things" (1978c, 112). Meaning as diacritical difference, subject as it is to all manner of oppositions and displacements (paradigmatic or syntagmatic: as in cat/hat or cat/dog) does not precede the victim; nor is it even proper to or possessed by the victim – except as a signifier of the difference between meaning and non-meaning.

The victim functions like the work of art which in the developing course of culture will come to represent it and replace it (with any luck definitively; cf. Gans 1981, 194–6). Not any more than an art work for us today, the victim is not a simple object of observation by gazing subjects. Because of the power (falsely) sensed as radiating from it, the victim transmits a subjective dimension, a dimension of conscious or willing agency; for the victim is experienced as the subject of an ambivalent power of which it is the object, the mediator, the sign. The victim's power to fascinate its beholders, to bind (whence 'fasciare') them in a state which allows neither for forward or backward movement – not stasis but hypostasis – issues from the double bind affecting the object of desire as it says: take me/don't take me. All those surrounding the object reinforce this bind.

The victim's power to fascinate is one with its power to mediate signification. The victim is accordingly the *mana* of the system of cultural differences; its function, as described by Lévi-Strauss whom Derrida quotes, "'is to be opposed to the absence of signification, without entailing by itself any particular signification'" (1978, 290). It is for this very 'antinomial' character (*ibid.*), which indeed is the character of a written character, of a letter, such as the anomalous 'a' in *différance*, that Derrida likens *mana* to writing, as the mark or the representation of a meaning which is in no wise present in it. Derrida does not actually thematize writing at this point of his discussion, but the supplement, which is what writing represents in our culture: a signifier in excess of an original meaning which it travesties or perverts. This is what writing represents to Lévi-Strauss in the wake of Saussure and of Rousseau before him and this is why, as Derrida amply demonstrates in *Of Grammatology*, they condemn it. For Derrida to the contrary: "It could no doubt be demonstrated that this *ration supplémentaire* of signification is the origin of the *ratio* itself" (*ibid.*). For Derrida, writing represents the 'supplement of origin', the mark or excluded member in which language originates. It is in and as the dead letter, which is what writing represents for Rousseau, that we find the origin of language:

> The death of speech is therefore the horizon and origin of language. But an origin and a horizon which do not hold themselves at its exterior borders. As always, death, which is neither a present to come nor a present past, shapes the interior of speech, as its trace, its reserve, its interior and exterior *différance*: as its supplement (1976, 314).

It is not in the mark itself, which doesn't mean anything, but in its exclusion, as the exclusion of non-meaning, that we must locate the origin of language. This is where Girard locates the victim, from whose exclusion meaning is thereafter free to proliferate. The victim is the supplement of origin in which, that is: in whose expulsion, the origin is (re)constituted. We are not speaking here of an historical reconstitution, which Girard recognizes as impossible (1978c, 52; Hamerton-Kelly, 89), but of a generative principle, whose very erasure accounts for propery mythological reconstitutions.

It is in terms of the ritual imperative, the need to recreate the foundational murder from which the new born community derives and perpetuates its existence, that we can understand this proliferation of meaning:

We have no difficulty understanding ritual reproduction. Spurred on by sacred terror, and in order to continue to live under the sign of the reconciling victim, men endeavor to reproduce this sign and to represent it. This consists first of all in the search for victims who seem most likely to provoke the primordial epiphany. This is where we must situate the first signifying activity, which is always already definable, if you insisit, in terms of language or of writing. And the moment arrives when the original victim, instead of being signified by new victims, by all sorts of things, which always signify that victim even as they mask it, disguise it and misunderstand it more and more (1978c, 113).

Any sort of thing can substitute for the victim precisely because the victim is always already a substitute, a signifier, a mark, in Derridean terms, of a deferral. From substitute to substitute, from disguise to disguise, from mark to model which is already a mark, the ritual imperative is the place where appropriative mimesis leads to, cedes to, defers to ontological mimesis; it is where the representation of desire is wed to the desire for representation. As Girard has recently recapitulated: "The victim must be the first object of non-instinctual attention, and he or she provides a good starting point for the creation of sign systems because the ritual imperative consists in a demand for substitute victims, thus introducing the practice of substitution that is the basis of all symbolizations" (Hamerton-Kelly, 129). Ritual sacrifice is a supplement to divine presence to which, in the expulsion of the victim, it ever only defers, which it ever only displaces. Representation is preferred to presence as the representation of violence is preferred to violence.

It is in Derrida's description of the logic of the supplement in Rousseau that we find an apt description of the ritual imperative. Rousseau has described masturbation as a 'dangerous supplement' to sexual possession (1977, 108—9), though he confesses that *jouissance* would be the death of him. As Derrida puts it:

Are things not complicated enough? The symbolic is the immediate, presence is absence, the nondeferred is deferred, pleasure is the menace of death. But one stroke must be still added to this system, to this strange economy of the supplement. In a certain way, it was already legible. A terrifying menace, the supplement is also the first and surest protection; against that very menace. This is why it cannot be given up. And sexual auto-affection, that is auto-affection in general, neither begins nor ends with what one thinks can be circumscribed by the name of masturbation. This supplement has not only the power of *procuring* an absent present through its image; procuring it for us through the proxy [*procuration*] of the sign, its holds it at a distance and masters it. For this presence is at the same time desired and feared. The supplement transgresses and at the same time respects the interdict. This is what also permits writing as the supplement of speech; but already also the spoken word as writing in general. Its economy exposes and protects us at the same time according to the play of forces and of the differences of forces (1976, 154—155).

The parallel is not as fortuitous as it may seem at first glance: masturbation for Rousseau is the representation of the satisfaction of desire, whereby it communicates with his literary fictions towards which he is analogously ambivalent, as witnessed by the second preface to *La Nouvelle Héloïse*; sacrifice similarly represents to the community the forces which generate it, and whose untrammelled release threaten to dissolve it. In a similar way as evoked by Derrida's last sentence, sacrificial representation is always attended by dread concern for ritual purity, lest the play of forces representing and disguising originary violence spill over into that violence itself; lest the representation of sacrifice lead to a sacrificial crisis, the mimetic violence of all against all from which ritual sacrifice is intended to protect us. (Rousseau's own desire for ritual purity, as if intended to perfect a sacrifice of which he is the victim, will be taken up in a later essay.)

The sacred is perceived as present in the victim, though it only prevails in what the victim represents, in what represents the victim: the saving, peaceful, mimetic presence of the community to itself. And the self, the ego, or self-consciousness, as the consciousness of a self which desires (Gans 1981, 50) is not at the origin of representation, but its by-product. Desire, in a word, 'is' differance as described by Derrida in the essay bearing this improper, anarchical anti-concept as its title: the becoming space of time, the becoming time of space:

In order for it [what is called the present] to be, an interval must separate it from what it is not; but the interval that constitutes it in the present must also and by the same token, divide the present in itself, thus dividing, along with the present, everything that can be conceived on its basis, that is, every being — in particular, for our metaphysical language, the substance or subject. Constituting itself, dynamically dividing itself, this interval is what could be called *spacing*; time's becoming-spatial or space's becoming-temporal (*temporalizing*) (1982, 143).

The Girardian subject as homologously divided, as differance, as desire, is constituted in the space-time of deferred appropriation in which notions of presence and absence originate. As Derrida writes of it in *Of Grammatology*:

This means that differance makes the opposition of presence and absence possible. Without the possibility of differance, the desire of presence as such would not find its breathing space. That means by the same token that this desire carries in itself the destiny of its non-satisfaction. Differance produces what it forbids, makes possible the very thing that it makes impossible (1976, 143).

The antithetical structure of differance operates according to the structure (or as Derrida will later have it in *La Carte postale*, the 'stricture') of the double bind as we read of it in *Violence and the Sacred*: "Man and his desires thus perpetually transmit contradictory signals to one another. Neither model nor disciple really understands why one constantly thwarts the other because neither perceives that his desire has become the reflection of the other's" (1978a, 147). Thus Girard's originary scenario can accommodate a post-structuralist critique of origins because it posits that origin and diffferance are one, that representation is the by-product of an originary differance, of mimetic desire, rather than the representation of an originary presence, of an origin of any kind.

Everything in this scenario moves along the bias, by the very same 'oblique and perilous movement' by which Derrida styles the task of deconstruction (1976, 14); everything is at acute angles of which the victim is the hinge. Representation is destined to indirection, since it is not the representation of an object but — via the object — of another's desire. The object is the sign of another's desire, which is itself the mimetic sign, the repetition, the reproduction, the representation of another's desire, and so on *ad infinitum*, that is, to the seemingly infinite proliferation of cultural representations. For it is this structure which generates the seemingly endless supplementarity by which one object of desire substitutes, stands in for and supplants another.

Nothing (but desire) takes place at the sacred center, which is why anything can take its place. Thus Rousseau describes taking Thérèse Levasseur as his common law wife: she is a supplement to 'Maman,' Mme de Warens, who is herself a supplement to Rousseau's natural mother who died at birth, who is herself a supplement to nature — unless nature is a supplement to her:

A little intimacy with this excellent girl, a little reflection upon my situation, made me feel that, while thinking only of my pleasures, I had done much to promote my happiness. *To supply the place of* my extinguished ambition, I needed a lively sentiment which should *take complete possession of* [literally 'fill' — *remplit*] my heart. In a word, I needed a successor to mamma. As I should never live with her again, I wanted someone to live with her pupil, in whom I might find the simplicity and docility of heart which she had found in me. I felt it was necessary that the gentle tranquility of private and domestic life *should make up* to me for the loss of the brilliant career which I was renouncing. When I was quite alone, I felt a void in my heart, which it only needed another heart to *fill*. Destiny had deprived me of, or at least in part, alienated me from, that heart for which Nature had formed me. From that moment I was alone; for *with me it has always been everything or nothing. I found in Thérèse the substitute* [*supplément*] *that I needed.* (Derrida 1976, 157)

The mother signifies an irreplaceable locus of intimacy; that place is as readily filled as one signifier displaces another. Derrida's comment on this 'chain of supplements' not only 'deconstructs' Rousseau's declared desire for presence; it explains the cultural role of signs as they issue from desire:

Through this sequence of supplements a necessity is announced: that of an infinite chain, ineluctably multiplying the supplementary mediations that produce the sense of the very thing they defer: the mirage of the thing itself, of immediate presence, of originary perception. Immediacy is derived. That all begins with the intermediary is what is indeed "inconceivable [to reason]" (1976, 157).

"Il n'y eut jamais pour moi d'intermédiaire entre tout et rien": between everything and nothing, there is nothing — that is, there is the signifier whose function is to display the difference between everything and nothing, between the thing itself and all cultural mediations. Rousseau's critique of the signifier, of the letter, is destined to bring its central role into view.

> I call the basic truth of every genus those elements in it the existence of which cannot be proved. As regards both these primary truths and the attributes dependent on them the meaning of the name is assumed. The fact of their existence as regards the primary truths must be assumed; but it has to be proved of the remainder, the attributes. (Aristotle)

The experience of the sacred as divine presence is born of a maculated conception; it is illusory but it is efficacious. Or rather, it is efficacious because it is illusory. The links (or macula, as in chain mail) which bind a community as a religion are forged by abstention from the object of desire. If the etymology of 'religare' is to be believed, the ties that bind the community together are double, issuing from a double bind that assures the sacred its foundational role. The sacred isn't anything or anywhere, it is only the force of attraction and repulsion attending the object of a desire which is sufficiently intense to inspire another desire which in turn is sufficiently intense to render appropriation dangerous, dreadful. The object, the victim, is only a signifier of desire. The sacred isn't anything but this mortal opposition, the resemblance of desires, which is why anything, any object, can represent the sacred to the precise extent that the sacred does not exist except as the signifer of desire's ambivalence. Anything, even writing, especially writing as the mark of an absence, of an abstinence prior to any awareness of presence or absence. Whence the properly victimary status of writing for Rousseau, of whom Derrida cannily observes: "What he excluded more violently than another must of course have fascinated and tormented him more than another" (1976, 98).

In the beginning, then, we can say with Derrida, is the trace, the body of the victim being the trace of a violence which, being mimetic, does not originate with it or with anyone. The origin of this violence, improperly attributed to the victim whose destruction ends violence, is properly unrepresentable; it is only proper to the rivalry of combatants, to their mimetic conflict. The sacralization of the victim erases this non-origin — the non-original but ony repetitive, mimetic origin of violence — when it enshrines the victim as the origin of the community. What Derrida writes of the arche-trace thus describes the victim in Girard's scenario:

> The value of the transcendental arche [*archie*] must make its necessity felt before letting itself be erased. The concept of arche-trace must comply with both that necessity and that erasure. It is in fact contradictory and not acceptable within the logic of identity. The trace is not only the disappearance of origin — within the discourse that we sustain and according to the path that we follow it means that the origin did not even disappear, that it was never constituted except reciprocally by a nonorigin, the trace, which thus becomes the origin of the origin. From then on, to wrench the concept of the trace from the classical scheme, which would derive it from a presence or from an originary nontrace and which would make of it an empirical mark, one must indeed speak of an originary trace or archetrace. Yet we know that that concept destroys its name and that, if all begins with the trace, there is above all no originary trace (1976, 61).

The adverb 'reciprocally' translates *en retour*, which is something else again as it signifies the retroactive constitution of the origin by the trace. This is just how Girard describes the origin of the human community in its active and passive relation to the victim. In Girard's scenario, all begins with the sacred, which destroys its name in the same way as described by Derrida; in the same antithetical way that interested Freud, but more importantly, more decisively for the nascent human community, in the way it erases the traces of human violence.

As Girard observes in the context of a discussion of Derrida to which we shall return, "Western thought continues to function as the erasure [effacement] of traces. It is no longer the direct traces of foundational violence that are being expelled, but the traces of a first expulsion, of a second, or even of a third, of a fourth. In other words, we are dealing with traces of traces of traces, etc." (1978c, 73).

These traces are legible in the institutions which culture develops with the repetition, proliferation and complication of ritual sacrifice: animal husbandry as it issues from the need to keep animals to substitute for human victims (Girard 1978c, 77ff.); kingship as it divides, bifurcates the function of the victim as founder and savior of the community from the equally necessary destruction of the victim (Girard 1978c, 59ff.); theater as it represents a sacrificial crisis which it no longer comprehends or from whose comprehension it shrinks back (Girard 1978a, Chaps. 3, 5); not to mention literary criticism, as it begins with Aristotle's efforts to protect dramatic poetry from Plato's canny expulsion of it (Girard 1978a, 292—295); not to mention narrative literature in general as it begins with the epic hero's destruction of a monster, that is of an unnatural, inhuman, i.e. sacred violence whose saving power is nonetheless invested *en retour* in the figure, the fiction of the hero. Then there is the law, which accords to the city a monopoly on violence in quest of the right victim, the culpable origin of violence (Girard 1978a, Chap. 1). And the procedures of the law, as heir to sacrifice, do not fail to display a dramaturgical structure which issues in turn from and remains party to a more primitive liturgical imperative; it is precisely when proper juridical procedure resembles more the demand for ritual purity than for transcendent goals of justice that the law is, as it seems now, in trouble (cf. Denvir, 825—826). And, too, there is psychoanalysis, whose sacrificial dynamics as a profession historically replete with mystified expulsions, has been admirably scrutinized by François Roustang in *Dire Mastery: Discipleship from Freud to Lacan*. Virtually all cultural institutions, from sports (Routeau 1980) to salad bars (Serres 1980; Gans 1985), are traceable to the double agency of the victim, to the mystifying expulsion of violence in the sacred.

Among cultural institutions, those devoted to representation are especially high on the agenda I am prescribing for the human sciences. I have mentioned theater; film, both popular (massively devoted to expelling the monster, to chasing bad violence with good: to sacralizing violence) and high-cultural (Livingston), is obviously high on the list. And books. Not merely because our hugely bookish culture derives in part from comment on, preoccupation with and subsequent rivalry with the Bible, the good book, the God book (McKenna, 1978a, b), but also for reasons advanced by Derrida when he seeks to "put in question . . . the unity of the book, and the unity 'book' considered as a perfect totality, with all the implications of such a concept." "And you know", he continues, "that these implications concern the entirety of our culture, directly or indirectly" (1981, 13). "Le tout de notre culture, de près ou de loin": why has no one observed that the range, the scale of Derrida's detotalizing ambitions is in no way incommensurate with that of Girard's 'unifying theory'?

Derrida's deconstructive target is the book as an institution which confides in the capacity of signifiers to represent signifieds which are ultimately independent of them; he is aiming at the book as the exemplary institution of logocentrism. As he writes in *Of Grammatology*'s first chapter, 'The End of the Book and the Beginning of Writing':

> The idea of the book is the idea of a totality, finite or infinite, of the signifier; this totality of the signifier cannot be a totality, unless a totality constituted by the signified preexists it, supervises its inscriptions and its signs, and is independent of it in its ideality. The idea of the book, which always refers to a natural totality, is profoundly alien to the sense of writing. It is the encyclopedic protection of theology and of logocentrism against the disruption of writing, against its aphoristic energy, and, as I shall specify later, against difference in general. If I distinguish the text from the book, I shall say that the destruction of the book, as it is now under way in all domains, denudes the surface of the text. That necessary violence responds to a violence that was no less necessary (1976, 18).

Denuding the surface of the text, tracking the book's operations back to the expulsion of writing that institutes it, uncovering that erasure is the task of deconstruction as pursued across the writings of Rousseau (as of Husserl in *La Voix et le phénomène*, and of many another in various essays). The mention of violence here is not fortuitous, not hyperbolic. Derrida has the same understanding of its complex structure and of its sacrificial effects as Girard. As he says, "The structure of violence is complex and its possibility — writing — is no less so" (1976, 112).

Books are the building blocks of Western culture — which he rightly observes is fast becoming a planetary culture (1976, 3, 10) — and Derrida sometimes plays on the phonemic vicinity of tome and tomb in order to emphasize their monumental status; it is that of a funerary monument, such as a pyramid in which a body is encrypted. What is encrypted in the book is the letter, as is the letter 'a' in differance, which we pass over in silence when we read it aloud: "The *a* of *différance*, therefore, is not heard; it remains silent, secret, and discreet, like a tomb. It is a tomb that (provided one knows how to decipher its legend) is not far from signaling the death of the king" ('Différance' in Derrida, 1982, 4; cf. also 'Le Puits et la pyramide', *ibid.*). The legend of Oedipus, of Pentheus, as told by Girard (1978a, Chaps. 3, 5), the legend of Romulus as told by Michel Serres (*Rome: Le Livre des fondations*), is the one Derrida tells of the letter. The silence surrounding the letter, the space it marks prior to signification, lies at the origin of language (1976, 44); that is to say, for all intents and purposes, at the origin of culture as a system of differences and the representation of differences. That is what culture is: the institution of difference whereby language is possible, and which language makes possible. It is the letter that represents, indeed embodies, incarnates this institution, that of the sign in its arbitrary relation to a natural environment: "The very idea of institution — hence of the arbitrariness of the sign — is unthinkable before the possibility of writing and outside of its horizon. Quite simply, that is, outside of the horizon itself, outside the world as space of inscription, as the opening to the emission and to the spatial *distribution* of signs, to the *regulated play* of their differences, even if they are 'phonic'" (1976, 44).

It is for this "rupture with nature" (1976, 36) that writing in turn is accused by Rousseau, as by Plato before him ('Plato's Pharmacy' in *Dissemination*) and Saussure and Lévi-Strauss after him, of "the *forgetting* of a simple origin", which it displaces, supplements, usurps. "Writing, a mnemotechnic means, supplanting good memory, spontaneous memory, signifies forgetfulness" and it is for this "violence of forgetting" that writing is liable in turn to expulsion (1976, 36). What is expelled in and as writing is the 'double' of representation whose simple origin in nature it supplants. What is expelled with writing is the notion of a non-simple origin, of an originary parricide or fratricide; what is expelled is the violence of doubles, or violence itself. The structure of violence is complex because it is violence itself which expels another violence, the violence of doubles which it expels in order to (re)institute, *en retour*, a simple, non-violent origin. "There is not an origin, that is to say a simple origin" for Derrida (1976, 174) because the latter is only constituted by repeating the violence it expels, by dissimulating an originary violence; for it dissimulates its own violence in the violence it expels. The simple origin is accomplice to the non-simple origin, the violence of doubles that it expels. This 'complicity of origins' as the violence which dissimulates itself in the expulson of its other, its other self, its rival double, its true violent origin, the origin of its truth as violence, as constituted "by a desire for speech displacing its other and its double and working to reduce its difference" (1976, 56) — this is what Derrida calls 'arche-writing': "This complicity of origins may be called arche-writing. What is lost in that complicity is therefore the myth of the simplicity of origin. This myth is linked to the very concept of origin; to speech reciting the origin, to the myth of the origin and not only to myths of origin" (1976, 92). There is no simple origin because its recitation requires the erasure of non-simple origins and the erasure of that erasure. The structure of violence is complex because it is accomplice to its own erasure. The structure of this complexity might be styled as violance, to encode the trace of its difference from itself, within itself; to encode its originary dissimulation. This is the very function of sacrifice for Girard, as it erases the human origins of violence in the expulsion of the victim, in the sacralization of the victim whose divine violence dissimulates the community's own violence. The victim — as for Derrida, the letter — is the fold, *le pli* — also for Derrida the hinge, *la brisure* (1976, 65) — in this structure of violence doubles, in this com*pli*city of origins.

The fold or *pli* is just what structuralism does not see in its confidence in non-violent origins, in naturally good savages who are living "in a 'state of culture' whose *natural* goodness had not yet been degraded" (1976, 112) — the trade mark of this degradation being writing, as argued by Lévi-Strauss in *Tristes Tropiques* (Ch. XXVII: 'La Leçon d'écriture'). In reviewing Lévi-Strauss's efforts to elicit from Nambikwara children their proper names, which they are forbidden to reveal (Ch. XXVI), Derrida uncovers a scenario of violent origins which Girard has uncovered elsewhere in the anthropologist's structural reading of myth (1978a, Chap. 9; 1978b, Chaps. 8, 9). The anthropologist succeeds by coaxing one child to tattle on another, whereupon the others follow suit in reprisal, *en guise de représailles*: "From then on, it was very easy, although rather unscrupulous, to incite the children against each other and get to know all their names. After which, having created a certain atmosphere of complicity, I had little difficulty in getting them all to tell me the names of the adults" (Lévi-Strauss, 279;

in Derrida 1976, 111). The taboo affecting the proper name prohibits revealing an order of classification, "the inscription within a system of linguistico-social differences" (1976, 111); that is to say: culture itself in its difference and differentiating activity from nature.

It is here that Derrida observes that the "structure of violence is complex": for there is a first violence, "the originary violence of language": in separating nature from culture; a second violence, "reparatory, protective", which conceals and thereby confirms this arche-violence; a third violence, instigated by the anthropologist, which reveals the concealment along with the original rupture whose violence the concealment functioned to expel from consciousness. "It is on this tertiary level, that of the empirical consciousness, that the common concept of violence (the system of the moral law and of transgression), whose possibility remains yet unthought, should no doubt be situated" (1976, 112). Each level of violence repeats the former in its dissimulating structure, each is a *mise en abyme* of originary violence which culture functions to conceal — and not least in the person of the anthropologist. He acknowledges himself as the culprit, *le coupable* (1976, 112), in instigating the mimetic violence of this '*guerre des noms propres*', as Derrida calls it. How shall we regard him? As the god presiding over this war which he instigates, authorizes on the one hand — whereby he is comparable to the neo-divinity which Rousseau enjoys in an analogous scene from the *Rêveries* (IX), which Derrida cites without commentary and to which we return in a later essay; but also as the victim, as what the victim represents to culture — the excluded member, the origin of a violence it in nowise conceives as its own.

Girard, for his part, acknowledges the importance of Derridean deconstruction for his hypothesis, including its importance for an anthropology whose possibility it questions. In a sense this essay is only intended to serve as a *mise au point*, a textual validation of these remarks by Girard:

> The discovery of the emissary victim as the mechanism of symbolicity itself justifies the discourse of deconstruction at the same time as it completes it. And this discovery explains at the same time the characteristic traits of this contemporary discourse. Because it hasn't yet managed to root itself in an anthropology of the emissary victim, it remains devoted to verbal acrobatics which are finally sterile. This is not for lack of the right words; it is only too talented where words are concerned; it is the mechanism behind the words which escapes this discourse. If you examine the terms which serve as

> the mainspring in Derrida's best analyses, you will see that is always a question, beyond the philosophical paradoxes being deconstructed, of the paradoxes of the sacred; it is never a question of deconstructing this, which sparkles all the more on that account in the eyes of the reader (1978c, 72).

I have resisted the temptation to elide the derogatory tenor of this passage, which goes on to associate deconstruction with literature in a pejorative, i.e. trivial, inconsequential sense — the sense in which Rousseau uses this word throughout his *Confessions*. Girard in turn has been rebuked for slighting literature (Pachet in Dumouchel, 1985, 385—394, and discussion, 408—417). To pursue this point as it richly deserves is not the aim of the present chapter, except to beckon attention to Eric Gans's theory (1985) of the origin of literature in resentment as Nietzsche analyzed it in the *Genealogy of Morals* — and from which he was far from immune (Girard, 1978b, Chap. 4). I think Gans's theory is born out, as ontogeny by phylogeny, by the atmosphere of mutual, i.e. mimetic, defensiveness pervading our contemporary critical environment. It is the sort of thing, for instance, which prevents Girardians and Derrideans alike from reading each other sympathetically, indeed from perceiving the pathology of origins which informs the texts being examined here. Of course the very pathos of readership itself is partly to blame, as it is structured by motives of assent and dissent anterior to the text. In order to live down my own 'faux beau rôle' in this regard, I shall try to make my conclusions as outlandish as possible.

> . . . on ne pratique pas impunément la duplicité de pensée, la pensée (du) double. (Borch-Jacobsen)

The origin, far from being one and indivisible for Girard, is, to cite his only neologism, 'interdividual' (1978c, part 3) as it proceeds from the mimetic rivalry of doubles whose violence is erased in the expulsion/ sacralization of the victim; as it proceeds from a violence whose human origin is erased in the difference that divinity makes. Following Derrida, we have spelled this difference, that of the sacred, with an 'a', as differance, to represent the sacred as the difference of human violence from itself, within itself, and as the dissimulation of that difference. The victim is the originary trace or the arche-trace of this violence and it destroys its name as sacred: as holy and accursed. It destroys itself as value in the same way, in the same sense that arche-trace destroys itself as arche-. The victim has no value; all values derive from its expulsion in the same way that the letter has no value, though all linguistic value, all meaning as representation derives from its entombment, or what its entombment represents: ob-literation of originary violence.

The sacred destroys its name in the same way that difference, as from an origin or as selfsame identity different from others, self-destructs as differance. As I mentioned earlier, it destroys its name in precisely the antithetical way that intrigued the author of *Traumdeutung*. For Freud this antithetical sense was important for understanding the language of dreams (Vol. XI, 162), important for a science of dreams which are the representation of wish fulfilment. Dreams, he declared, are the 'royal road' to the unconscious, the path we have to trace to formulate the science of desire that goes by the name of psychoanalysis. In *Le Sujet freudien*, Mikkel Borch-Jacobsen has argued extensively — and I do not hesitate to say definitively — that Freud's own desire for originality blinded him to the originality of his own discovery about dreams; that in his rivalry with Jung for scientific precedence and ascendancy, he did not sufficiently attend to the mimetic origin of desire that is encoded, distorted, misrepresented in dreams. Borch-Jacobsen's stunning analyses have the further, not to say decisive interest from the viewpoint of this essay of being a reading of Freud which combines, indeed fairly synthesizes Girardian and Derridean perspectives. Perhaps the only thing the author does not do, leaving it for this easy to do, is to thematize his own hermeneutic achievement, his own critical gesture; to thematize, that is, the remarkable correlations between Girard and Derrida — so as to argue, for instance, for an agenda in the human sciences that Derrideans, perhaps more than Derrida himself, are apt to dismiss.

For nothing Derrida advances by way of his critique of origins, of a non-simple origin, or by way of his critique of representation as the representation of an original presence, nothing that goes by the name-which-erases-its-name of supplement, arche-trace, differance, etc. (or by another name, 'pharmakos', as I have argued elsewhere [cf. McKenna, 1991, Chap. 2]) — is proof against Girard's anthropological hypothesis of human origins as rooted in the dynamics of mimetic desire. *Au contraire*: a 'principled' critique of structuralism, a deconstruction of ontotheology, or ontotheologocentrism as it is sometimes monstrously labeled, is one that clears a path to a genetic hypothesis that postulates the 'complex' and accordingly undecidable origins of desire and the sacred, of the human and the holy as they issue in turn from a relation to a crypt, a corpse, a dead letter, a pure signifier.

The origin, or at the origin, is a repetition for Derrida. Girard's hypothesis argues no less. Paraphrasing some verse of Edmund Jabès, whose victimary thematics as expatriate Egyptian Jew preoccupied with books and holocaust fairly predestine him a place in Derrida's canon, the latter writes:

> Death is at the dawn because everything has begun with repetition. Once the center or the origin have begun by repeating themselves, by redoubling themselves, the double did not only add itself to the simple. It divided it and supplemented it. There was immediately a double origin plus its repetition. Three is the first figure of repetition. The last too, for the abyss of representation always remains dominated by its rhythm, infinitely. The infinite is doubtless neither one, nor empty, nor innumerable. It is of a ternary essence (1978, 299).

These are not merely beguiling paradoxes; they repeat the structure of the scene(s) we have just reviewed, the scene of the foundational murder (Girard), the originary scene of representation (Gans), which Borch-Jacobsen (149) incisively labels the "*Autre scène comme outre-scène, avant scène de la mimesis originaire*", and of which the letter, as a repetition, is the mark, the model. Three, not two, is the original figure of repetition for the same reason that it is the ultimate figure of infinity, and of circularity, of eternal return, of *Wiederholungszwang*, the compulsion to repeat: because there is no object for a subject, as there is no nature for a culture, prior to another subject which designates it as an object of desire and that subject in turn is only constituted by the other subject's desire, and so on infinitely, undecidably. For the origin of desire is undecidable, being ever only, or 'always already', the copy, the repetition of another desire. It is always another subject which institutes a subject in its 'undecidable identity' (Borch-Jacobsen, 65). It is in virtue of this 'recurrence that is originary' that the subject is forever *en retard sur son origine*: "The other was always already at the origin (without origin) of the individual and everything had already begun — everything, that is to say: history — in a *pre*history that is *pre*individual" (*ibid.* 147, 286).

Desire is the origin of this undecidability as it bears on a victim whose attraction assures its repulsion — whereby it is also true that undecidability, as attraction-repulsion, is the origin of desire. These two statements do not contradict each other; rather they are themselves undecidable. In this sense, it is true, as a Marxist Derridean (or a Derridean Marxist, the critical impetus being undecidable) has stated, that "Undecidability *is* truth" (Ryan, 7). But this is not grounds for conclusions about our inability to know anything, as some Derrideans (not Ryan) have argued. On the contrary, it is grounds for understanding desire as what makes us human, and never more so than when we dissimulate its origin and its power by locating them in the object. It is also grounds for understanding our desire for truth, for understanding what makes truth desirable: the victim.

The victim is the origin of an indecision, the focus of an originary indecision, of an irreducible ambivalence, in which our species is born along with the sacred. Desire is, if you will, the origin of the sacred, which is only the erasure of the mimetic (non)origin of desire; the sacred is the originary *dissimulation* of violence in the original sense of the word, which is that of concealing a resemblance. The logic of desire is contrary to the logic of identity and difference, of identity and non-contradiction, both as to its origin and as to its signification. A culture develops according to principles of identity and difference as long as it succeeds in masking that resemblance. It develops in differentiating between good and bad victims as between others and one's self, one's own.

Today we apprehend that culture is in crisis (as perhaps all great writers have seen: "*Qui l'interroge dans son principe l'anéantit*", said Pascal), and not least because of activities like deconstruction, which we may regard as "a new name for some old ways of thinking". This, of course, is the way William James subtitled Pragmatism, whose alliance with Derrida has been suggested by Richard Rorty in *Consequences of Pragmatism* (90—109) — though the full consequences of this 'post-philosophical' juncture await elaboration. This way is at least as old as the Enlightenment in its detection of false dichotomies, of illusory differences, such as man and Moslem, human and Hottentot with a Montesquieu, a Voltaire, a Diderot.

These days only one difference remains and it is irreducible, or 'uncircumventable', 'unbypassable', as the French word *incontournable* is variously translated from Derrida. It is a good word; it means you cannot get around it. It is the difference between victim and persecutor, in the name of which the quest for truth as opposed to sacred or sacralized authority is decided. These decisions are being made everyday in our institutions; in our schools, our law courts and councils of power, these decisions are increasingly called into question as the difference between good and bad victims becomes increasingly problematic, as revealing traces of violence which all disown; which all disown as one's own violence, for we are always prepared to confront another's violence as violently as you please. As Girard observes, it is victims themselves that we hurl at each other in our denunciations of violence (Hamerton-Kelly, 140). Disowning one's own violence as another's, as being the other's, as originating with the other, is the surest means of its perpetuation. It is the functional, effective — and not merely formal — equivalent of dissimulating the other's desire as our own, of claiming as our own the other's desire.

So the wheel goes round, but not, decidedly not, infinitely. In the age of nuclear proliferation — or in Derridean terms: 'dissemination', which can be shown to have the same aboriginal, anarcheological structure as both nuclear fission and strategic deterrence (McKenna, 1991, Chap. 4) — the victim is the focus of a decision from which mankind cannot distract its attention, since the wrong decision would prove definitive for the species.

So we have deterrence, which we would also do well to spell with an 'a', as deterrance, for at least two reasons: it invites us to recall the terrain it protects, and to recall the terror (from 'de-terrere') in which it originates. For deterrance is nothing but the deference we pay to the sacred, to absolute violence, in its difference from itself, or its differance as it originates in mimetic dissimulation. No one is immune to that violence, which Winston Churchill was perhaps the first to evoke in a language replete with the dynamics of our sacred origins: "Safety", he said to the House of Commons in 1955, "will be the sturdy child of terror, and survival the twin brother of annihilation" (cited in Schell, 197). As Girard wrote of the sacrificial god, Dionysus by name, but Titan, Saturn, Minuteman or any other will do, "He will be as benevolent from afar as he was terrible in propinquity" (1978a, 134). Derridean analysis helps us to see how the Girardian sacred is the origin of propinquity and of distance, of presence and absence, of a time before and a time after, etc. Girardian analysis protects us in turn from the delusion that originary differance propels or destines us to infinite meaninglessness or to complacency about our finitude. Indeed Girard lends a meaning and a value to deconstruction which its adherents seem only too anxious at times to elide or ignore.

We have grown accustomed to thinking that we inhabit a world in which absolute values are absent, and have been absent since the French Revolution or the Enlightenment, or since Cartesian epistemology or since Galileo — whose status, by the way, as a victim of sacralized persecution accounts as much as anything else for the destabilizing import of his scientific discovery: *The Crime of Galileo*, his trial as narrated by Giorgio de Santillana, has all the marks of a "botched kill," which is how Girard leads us to see the trials of Job (1985, 42). Anyway, all our presumed relativism is nonsense. We inhabit a world of absolute values. For proof we need only regard the violence which attends the imposition of or the rivalry between these values. Violence is nothing if it is not the signifier of the absolute, the form and substance of absolute desire. The absolute is alive and well among us; the question is whether we can live it down, or outlive it, survive it. It is a question of 'Living on', 'Survivre', to borrow from the title of another essay by Derrida which I discussed elsewhere in this context (1991, Chap. 4). The indecision attending this question is a scandal to our ordinary powers of rational deduction. Yet it should not surprise us if we read Girard correctly, for our desire to live on hinges on decisions which desire, being what it is (the other's, allogenetic), does not make.

It is Girard's conception of desire, as revealed by our greatest writers according to *Deceit, Desire and the Novel*, which accounts for the seemingly derealizing effects of deconstruction as it undermines our confidence in such concepts as subject and object, substance, referent, etc. — the entire host of metaphysically derived concepts which we take to be percepts, realities. In the Derridean critique of ontology, these are to be inscribed within an economy of differance:

> Nothing — no present and in-*different* being — thus precedes *différance* and spacing. There is no subject who is agent, author, and master of *différance*, who eventually and empirically would be overtaken by *différance* [which *différance* would take the place of, 'surviendrait']. Subjectivity — like objectivity — is an effect of *différance*, an effect inscribed in a system of *différance*. This is why the *a* of *différance* also recalls that spacing is temporalizing, the detour and postponement by means of which intuition, perception, consummation — in a word, the relationship to the present, the reference to a present reality, to a *being* — are always *deferred*. Deferred by virtue of the very principle of difference which holds that an element functions and signifies, takes on or conveys meaning, only by referring to another past or future element in an economy of traces. This economic aspect of *différance*, which brings into play a certain not conscious calculation in a field of forces, is inseparable from the more narrowly semiotic aspect of *différance*. It confirms that the subject, and first of all the conscious and speaking subject, depends upon the system of differences and the movement of *différance*, that the subject is not present, nor above all present to itself before *différance*, that the subject is constituted only in being divided from itself, in becoming space, in temporalizing, in deferral; and it confirms that, as Saussure said, "language [which consists only of differences (JD)] is not a function of the speaking subject" (1981, 28–29).

Indeed not, since language is a function of desire precisely as differance. The subject's 'being divided by itself' is its mimetic constitution, or what Girard calls its interdividuality. Once we conceive differance as desire, whose objects are but the signifiers, the traces of another desire, we can understand the ontologically decentering import of Girard's displacement of both subject and object by desire 'itself':

> If desire is the same for all men, if there is only one and the same desire, there is no reason not to make of it the true 'subject' of the structure, a subject moreover which leads back to mimesis. I avoid saying 'the desiring subject' in order not to give the impression of falling back into a psychology of the subject (1978c, 327).

As there is for Derrida "no economy without *différance*," which is "the most general structure of economy" (1981, 8), there is for Girard no economy — not even in the 'restricted' sense, the one Derrida calls 'classical' — without desire. In order to fully demonstrate this homology, we need to relate Derrida's discussion of Georges Bataille, entitled 'De l'économie restreinte à l'économie générale' (1978), to the several works of a decisively Girardian stamp devoted to this topic by Georges-Hubert Radkowski (*Les Jeux du désir: De la technique à l'économie*), by Michel Aglietta and André Orléan (*La Violence de la monnaie*), by Paul Dumouchel and Jean-Pierre Dupuy (*L'Enfer des choses: René Girard et la logique de l'économie*), by Eric Gans (*The End of Culture: Towards a Generative Anthropology*, which acknowledges the economic import of differance, while further engaging the socioeconomic theories of Marvin Harris, Marshall Sahlins and Jean Baudrillard). Pending that agenda, suffice it for the nonce to observe that according to Girard's conception, everything passes by the detour and postponement of desire, which functions and signifies like the elements described in Saussurian linguistics.

> There is a definite social relation between men, that assumes, in their eyes, the fantastic form of a relation between things. In order, therefore, to find an analogy, we must have recourse to the mist-enveloped regions of the religious world. (Marx)

> If we look at the most enlightened portion of the world, we see the various States armed to the teech, sharpening their weapons in time of peace the one against the other. (Kant)

Economic competition is ruled by mimetic desire, which is not ruled, governed or mastered in any way by a value emanating from the object, the commodity, but only by another desire. The advertising industry exploits this fact when it regularly focuses less on the object of consumption than on the image, i.e. imitation, of its hierophantically self-possessed consumers. Thus too, from the present perspective, Marx's genially comic analysis of the commodity fetish approaches its apex when it describes the "metaphysical subtleties", the "theological niceties", as a "necromancy that surrounds the products of labor as long as they take the form of commodities" (*Capital* 71—76). For these and still other religious figures express the sacralization of the product in its "social form", whereby "the process of production has the mastery over man, instead of being controlled by him" (81). Necromancy especially suggests a corpse whose interpretation governs the economy, which for Marx means that something, a product of labor, is transformed into nothing, "a social hieroglyphic" (74). Notwithstanding Marx's efforts to revive that corpse by demystifying its producers, what is of interest for us here is the simultaneous religious inflation and ontological nullity of the object in its social form. The two always go together; indeed they reinforce each other as the mimetic nature of competition and conflict works to assure "son néant ultime d'objet" (Girard, 1978c, 40)

> Nothing is more difficult than to admit the fundamental nullity of human conflict. For the conflicts of others, no problem; but for our own conflicts, it is almost impossible. *All* modern ideologies are huge machines which function to justify and legitimate just those conflicts which these days could put an end to mankind's existence. Man's entire madness is right there. If we don't admit the madness of human conflict today, we will never admit it. If the conflict is mimetic, the equally mimetic resolution leaves no residue; it purges the community entirely precisely because *there is no object* (1978c, 40; cf. also 334—35).

No more than the subject does the object take precedence in this *Discours sur le peu de réalité*, which is how Girard, taking the title from the Surrealists, styles the genius of a Flaubert whose *Bouvard et Pécuchet* attacks the "science and ideology — the very essence of the bourgeois conception of reality all-powerful at the time" (1966, 151; on Flaubert's vision of Holocaust, cf. McKenna, 1982).

What takes precedence to subject and object alike, as to all ontological determinations, is desire, precisely as the desire for precedence. It is still ontological desire if you will, if we conceive it as a desire for being which the other unknowingly transmits to us, with which the other con*tam*inates us (according to the undecidable structure of the 'entame' in Derrida):

> Once his basic needs are satisfied (indeed, sometimes even before), man is subject to intense desires, though he may not know precisely for what. The reason is that he desires *being*, something he himself lacks and which some other person seems to possess. The subject thus looks to that other person to inform him of what he should desire in order to acquire that being. If the model, who is apparently already endowed with superior being, desires some object, that object must surely be capable of conferring an even greater plenitude of being (1978a, 146).

The terms of ontological plenitude, of ontotheology as critiqued by Derrida, are in place; what is missing in Derrida's critique is any focus on the mimetic rivalry which generates these terms, these conceptual illusions. As this rivalry intensifies, violence itself becomes the object of desire (Girard 1978a, 145), displacing that object, the other's object or the other's very being as an object of desire — until things heat up to such a degree of intensity that violence will displace desire altogether as subject. This is where we find ourselves today, as we contemplate the rivalry for the disposition of total violence, of "total and remainderless destruction", as Derrida evokes it in his contribution to 'nuclear criticism' (1984, 27).

But the prospect of total war was always already *present* with the dual emergence of subjects, of mimetic doubles. It is the original and ultimate destiny, the eschatology of our ontological project, as Mikkel Borch-Jacobsen demonstrates by way of this Girardian paraphrase of Hegel (or Hegelian paraphrase of Girard, for what matters the originating authority?):

> The rage which takes hold of me at the sight of my neighbor does not come from the fact that he dispossesses me of something, but rather that he steals me, inexplicably, from myself. Whence the total, totalitarian character of the war in which we are engaged and which no satisfaction can pacify: 'It's him or me.' . . . There is a violence inherent to the very *appearance* of the other; all empirical violence testifies to it; it has no other 'reason' than desire (*that is to say self-consciousness*) which seeks its own self in the other and wishes to exist for its own self, independently and close by itself in its self-possession [dans sa propriété: as its own property, etc.]. Desire is violence because it is desire to be one's own self, [*désir d'être-propre*], desire of propriation and, as such, a desire that is allergic, murderous (115).

For Borch-Jacobsen, this *désir d'être-propre* engages all the ambiguities of property, of authorship and ownership, that Derrida explores in 'White Mythology' as undecidable (*Marges*). "*Le désir est un désir impropre du propre* — mimesis de propriation" (115).

This was doesn't happen every day, not ostensibly so. We have institutions whose exact purpose is to mask the violence informing relations between self and other, to deflect it away from the community. We know that they are breaking down, that they cannot withstand the de-institutionalizing, deritualizing impulses of Western culture, which is perhaps that of all culture (Gans, 1981, 227), and which we have called Enlightenment and which today goes by the name of deconstruction, and which Nietzsche, anticipating Girard on this point among others, ascribed to the relentless truth-seeking impulse of Christianity itself:

*What*, in all strictness, has really *conquered* the Christian God? The answer may be found in my *Gay Science* (section 357): "Christian morality itself, the concept of truthfulness taken more and more strictly, the confessional subtlety of the Christian conscience translated and sublimated into the scientific conscience, into intellectual cleanliness at any price" (*Genealogy*, part III, section 27).

The refined linguistic sensibilities of deconstruction reflect this impulse even as it attacks our scientific conscience, as when it archly deconstructs Enlightenment. Where Girard differs from it, as from Nietzsche as well, is in his focus on concern for the victim as the origin of our truth-seeking, and as revealed in Scripture (1978c, part 2; Hamerton-Kelly, 109—145; Girard 1982, Chap. 15; 1985 *passim*).

According to this view, the breakdown of institutional Christianity itself is the legacy of the crucifixion narrative which is one with the Hebrew Bible's denunciation of overtly sacrificial institutions. In our day the denunciation extends to the covertly sacrificial institutions, such as the law (Unger; Denvir; Norris, Chap. 7), but also war itself, which in the nuclear age is "*fabulously textual*, through and through" (Derrida, 1984, 23) for being unwinnable, unwageable, unthinkable — or at least *inénarrable* for leaving no witnesses to represent it. For Derrida, the ontological implications are unambiguous as it finally resolves our crisis of representation:

The only referent that is absolutely real is thus of the scope or dimension of an absolute nuclear catastrophe that would irreversibly destroy the entire archive and all symbolic capacity, would destroy the 'movement of survival', what I call '*survivance*', at the very heart of life. This absolute referent of all possible literature is on a par with the absolute effacement of any possible trace; it is thus the only ineffaceable trace, it is so as the trace of what is entirely other, '*trace du tout autre*'. This is the only absolute trace, — effaceable, ineffaceable. The only 'subject' of all possible literature, of all possible criticism, its only ultimate and a-symbolic referent, unsymbolizable, even unsignifiable; this is, if not the nuclear age, if not the nuclear catastrophe, at least that towards which nuclear discourse and the nuclear symbolic are *still beckoning*: the remainderless and a-symbolic destruction of literature. Literature and literary criticism cannot speak of anything else, they can have no other ultimate referent, they can only multiply their strategic maneuvers in order to assimilate that unassimilable wholly other (1984, 28).

Derrida does not name this 'wholly other' as the sacred; he doesn't have to, for that is what all the philosophers since Hegel have named it. But Derrida exceeds the imaginative reach of philosophers in naming its referent as absolute violence. If it is the 'subject' of all literature and criticism, as of all possible cultural discourse, it is because violence, as Girard argues, is the true subject, the active and passive agent, the agendum and dilendum, of all cultural institutions, of institutional formation and dissolution. In answer to the question about the seemingly 'hypostatic', personified role of violence in his discourse, he asserts that

. . . violence, in all cultural orders, is always in the final analysis the true *subject* of every structure that is ritual, institutional, etc. From the time when the sacrificial order starts to decompose, this subject can only be the *Adversary* par excellence to the institution of the Kingdom of God. This subject is the devil named by tradition, precisely the one whom theology declares to be the subject and whom it nonetheless declares not to exist (1978c, 233).

The duality of the devil is redoubled in itself, being that of the Antichrist, the traditional rival double of the savior, and that of a being whose power issues from its not being, as being only the illusory being of the other. Girard reverses the theological conundrum that declares one of the ruses of the devil to be to persuade us that he does not exist, to persuade us accordingly that evil does not exist, whereby such duplicity the sway of evil among us is assured the more. For Girard, the ruse of the devil is to persuade us that he does exist in person, whereby the evil men do is expelled, projected outside them, into the agency of an incorrigibly ill will which would be the alien rival of man's good will. But to so personify the devil is to mask the role of the mimetic double, of the violent rival of a will which is ill for being but the transposition or transfer of the other's desire. The devil does not exist except as this *qui pro quo* of mimetically violent doubles, except as this rivalry of non-entities (for more on this double talk, cf. McKenna, 1983). Its definitive role in history, its role as the absolute subject and the absolute referent of history is nonetheless assured as long as we cling to this ontological delusion.

In the end is the double, as 'it' 'was' from the very beginning when, as the story goes, man first ate of the apple to be like gods in the knowledge of good and evil. Today we have this knowledge as never before — unless, as Girard urges, we include the likes of Cervantes, Dostoevsky, Shakespeare and Proust in the ranks of our human scientists: "Only a pseudo-science runs counter to the greatest works of our literary heritage. A real science will justify their vision and confirm their superiority" (1978b, 78). It is absolute knowledge, being knowledge of the absolute, even as Derrida allows. All the rest is literature. This is knowledge which culture has to this day depended upon not knowing in order to survive, as Girard writes of our sacrificial crisis:

Violence will come to an end only after it has had the last word and that word has been accepted as divine. The meaning of this word must remain hidden, the mechanism of unanimity remain concealed. For religion protects man as long as its ultimate foundations are not revealed. To drive the monster from its secret lair is to risk loosing it on mankind. To remove men's ignorance is only to risk exposing them to an even greater peril. The only barrier against human violence is raised on this misconception. In fact, the sacrificial crisis is simply another form of that knowledge which grows greater as the reciprocal violence grows more intense but which never leads to the whole truth (1978a, 135).

Well, now we know the truth and no radical critique of origins can dissuade us of it. For violence not to have the last word may depend upon our understanding the first word, the *Überworte*, as it names a victim whose place is taken by the sacred in order to mark a divine origin which never took place. This cannot go on, there being no place for violence if absolutely everyone is to be its victim. This is not merely a moral determination, as when we say "there is no place for that sort of behavior here". It is also an ontological determination, there being literally no place for violence to exercise its sovereignty between subjects, however deluded as to their origin, who no longer exist.

*Loyola University of Chicago*

REFERENCES

Aglietta, Michel and Orléan, André: 1982, *La Violence de la monnaie*. Paris: PUF.
Benveniste, Emile: 1969, *Le Vocabulaire des institutions indo-européennes*. Vol. 2: Pouvoir, droit, religion. Paris: Minuit.
Borch-Jacobsen, Mikkel: 1982, *Le Sujet freudien*. Paris: Flammarion.
Denvir, John: 1987, 'William Shakespeare and the Jurisprudence of Comedy'. *Stanford Law Review*. XXXIX: 4.

Derrida, Jacques: 1967, *La Voix et le phénomène: Introduction au problème du signe dans la phénoménologie de Husserl*. Paris: PUF.

Derrida, Jacques: 1972, *La Dissémination*. Paris: Seuil.

Derrida, Jacques: 1976, *Of Grammatology*. Trans. G. Spivak. Baltimore: John Hopkins UP.

Derrida, Jacques: 1978, *Writing and Difference*. Trans. Alan Bass. Chicago: U of Chicago P.

Derrida, Jacques: 1979, 'Living On' in *Deconstruction and Criticism*. Ed. H. Bloom. New York: Seabury.

Derrida, Jacques: 1980, *La Carte Postale: De Socrate à Freud et au-delà*. Paris: Flammarion.

Derrida, Jacques: 1981, *Positions*. Trans. Alan Bass. Chicago: U of Chicago P.

Derrida, Jacques: 1982, *Marges de la Philosophie*. Trans. Alan Bass. Chicago: U of Chicago P.

Derrida, Jacques: 1984, 'No Apocalypse, not now (Full speed ahead, seven missiles, seven missives)'. Trans. Catherine Porter and Philip Lewis. *Diacritics*. XIV: 2.

Dumouchel, Paul and Dupuy, Jean-Pierre: 1979, *L'Enfer des choses: René Girard et la logique de l'économie*. Paris: Seuil.

Dumouchel, Paul (ed.): 1985, *Violence et vérité: Autour de René Girard*. Paris: Grasset.

Freud, Sigmund: 1953—74, *The Standard Edition of the Complete Psychological Works of Sigmund Freud*. ed. James Strachey et al. London: Hogarth.

Gans, Eric: 1981, *The Origin of Language: A Formal Theory of Representation*. Berkeley: U of California P.

Gans, Eric: 1985, *The End of Culture: Toward a Generative Anthropology*. Berkeley: U of California P.

Girard, René: 1966, *Deceit, Desire and the Novel: Self and Other in Literary Structure*. Trans. Yvonne Freccero. Baltimore: Johns Hopkins UP.

Girard, René: 1978a, *Violence and the Sacred*. Trans. Patrick Gregory. Baltimore: Johns Hopkins UP.

Girard, René: 1978b, *"To Double Business Bound"*: *Essays on Literature, Myth, Mimesis and Anthropology*. Baltimore: Johns Hopkins UP.

Girard, René: 1978c, *Des Choses cachées depuis la fondation du monde*. Paris: Grasset.

Girard, René: 1982, *Le Bouc émissaire*. Paris: Grasset.

Girard, René, 1985, *La Route antique des hommes pervers*. Paris: Grasset.

Robert G. Hamerton-Kelly (ed.): 1987, *Violent Origin: Walter Burkert, René Girard and Jonathan Z. Smith on Ritual Killing and Cultural Formation*. Stanford: Stanford UP.

Lévi-Strauss, Claude: 1973, *Tristes Tropiques*. Trans. Johns and Doreen Weightman. New York: Atheneum.

Livingston, Paisley: 1982, *Ingmar Bergman and the Rituals of Art*. Ithaca: Cornell UP.

Marx, Karl: 1974, *Capital*. Trans. Samuel Moore and Edward Aveling. New York: International Publishers.

McKenna, Andrew: 1978a, 'Biblioclasm: Joycing Jesus and Borges'. *Diacritics*. VIII: 2.

McKenna, Andrew: 1978b, 'Biblioclasm: Derrida and his Precursors'. *Visible Language*. XII: 3.

McKenna, Andrew: 1982, 'Allodidacticism: Flaubert One Hundred Years After'. *Yale French Studies*. 63.

McKenna, Andrew: 1983, 'Double Talk: Two Baudelairean Revolutions'. *New Orleans Review*. X: 4.

McKenna, Andrew: 1991, *Violence and Difference: Girard, Derrida, and Deconstruction*. Champaign, IL: U of Illinois P.

Nietzsche, Friedrich: 1967, *On the Genealogy of Morals*. Trans. Walter Kaufmann. New York: Vintage.

Norris, Christopher: 1985, *The Contest of Faculties: Philosophy and Theory after Deconstruction*. London: Methuen.

de Radkowski, Georges-Hubert: 1980, *Les Jeux du désir: de la technique à l'économie*. Paris: PUF.

Rorty, Richard: 1982, *Consequences of Pragmatism*. Minnesota: U of Minnesota P.

Rousseau, Jean-Jacques: 1964, *Oeuvres complètes*. Eds. Bernard Gagnebin and Marcel Raymond. Paris: Gallimard, "Pléiade." 3 vols.

Rousseau, Jean-Jacques: 1977, *Confessions*. Trans. J. M. Cohen. New York: Penguin.

Roustang, François: 1982, *Dire Mastery: Discipleship from Freud to Lacan*. Trans. Ned Lukacher. Baltimore: Johns Hopkins UP.

Routeau, Luc: 1980, 'Au Cirque'. *Esprit* (May).

Ryan, Michael: 1982, *Marxism and Deconstruction: A Critical Articulation*. Baltimore: Johns Hopkins UP.

Santillana, Giorgio: 1955, *The Crime of Galileo*. Chicago: U of Chicago P.

Schell, Jonathan: 1982, *The Fate of the Earth*. New York: Knopf.

Serres, Michel: 1980, *Le Parasite*. Paris: Grasset.

Serres, Michel: 1983, *Rome: Le Livre des fondations*. Paris: Grasset.

Unger, Roberto: 1984, *Passion: An Essay on Personality*. New York: Free Press.

PAUL DUMOUCHEL

# A MORPHOGENETIC HYPOTHESIS ON THE CLOSURE OF POST-STRUCTURALISM

Professor McKenna has offered us an attempted conciliation or reconciliation of Girard and Derrida. In his own words his paper sets out "to relate Derrida's critique of origins to Girard's originary hypothesis" (p. 46).[1] Clearly, as the text unfolds, the reader sees that he is not dealing with an essay on the origin of social order and its relationship to violence but with an essay on Girard and Derrida and their relationship, or rather the relationship between their works. True enough, McKenna's paper is about violence, but it is not so much about the original violence that founds the social order according to Girard, as it is about another violence, final, terminal, definitive. 'Apocalypse' as the title of his paper says, 'total and remainderless destruction' as Derrida says (p. 71), nuclear extermination. In short, it is about a violence that founds nothing at all, out of which no social order can emerge. McKenna concludes that this final violence echoes the original violence from which the community was born, and that this resonance transforms, in a sense, the originary hypothesis into "absolute knowledge, being knowledge of the absolute" (p. 74). The absolute being total annihiliation.

Now there is nothing wrong in writing a paper about Girard and Derrida, after all they both are very interesting authors. And there is nothing wrong in writing a paper which attempts to reconcile them, even in a work on violence and the origin of social order. Yet it is somewhat strange that the introductory paper should deal so little with the theme of this part of the book: Violence and the Origin of Social Order.

In what follows I will suggest that this is not an accident, but a necessary consequence of Professor McKenna's endeavour to reconcile Girard and Derrida. It should be clear from the start that I think that McKenna's attempt is very successful. In other words, I believe that he has discovered real similarities between Girard and Derrida, and I find him very convincing, very efficient in making those similarities visible or palatable to his readers. My claim thus is not that these similarities are not real, but rather that these resemblances between our two



*Francisco J. Varela and Jean-Pierre Dupuy (eds), Understanding Origins*, 77—90.

authors, between the "contradictory logic of the sacred" and the "contradictory logic of the supplement" (p. 46), because they are embedded in contradictory conceptual frameworks reveal entirely different points of view on the question that interests us. My claim in short is that in Derrida we cannot speak of the origin of social order, or of anything else for that manner, and that this is why Professor McKenna has so little to say about it.

My paper will thus proceed in the following way: first of all I will recall and summarize Girard's originary hypothesis, then I will discuss Derrida's notions of origin and sign, and finally I will try to show that implicit in Derrida's texts is a notion of origin which cannot accomodate his own criticism of origin.

Girard proposes a theory of the origin of the social order. A theory that says, as McKenna reminded us, that in the beginning there is violence, undifferentiated violence, undifferentiating violence. In the beginning is the mimetic crisis, the war of all against all. A violence that is without reason, a war that has no object. That is, perhaps, the fundamental discovery, the essential claim of Girard, that at the origin of violent opposition there is no object, that strife and conflicts are without object in every sense of the word. The roots of enmity are not to be found in scarcity, and at the origin of violence there is no precious prize. Of course, all the fighters believe that at the end of the battle there is a valuable lot to be won; Girard tells us that the prize is worthless, that the treasure box is empty. This is not mere moral wisdom, something we should all know, and that would make us all immensely more happy if we knew it, and acted upon it. No, this is also knowledge, science, says Girard, a theory that is to be judged on its explanatory power.

If in the beginning there is violence, at the origin of violence, that is in the beginning of the beginning, there is mimesis, or mimetism, imitation. As Aristotle first noticed, "man differs from other animals in that he is more prone to imitation".[2] The idea is simple. We imitate each other and this proclivity is without bounds; imitation should not be limited to behaviors of representation: they dress the same way, they walk the same way. To the contrary, imitation should be, by hypothesis, allowed to bear on any and every type of behavior, including behaviors of appropriation, the various gestures we make to take, collect, grasp, accumulate, catch, capture, get. Once that is done, it is quite clear that imitation, mimesis, is no more the pale engine of social conformity which it is so often believed to be, though it is that also. For if we imitate each other in the gestures we make to gather and to possess objects, then we will soon become obstacles to each other in our common drive towards the same object. That, according to Girard, is the source of all conflicts, the origin of violence. Violence is without object because the object upon which the opponents converge is not, in spite of what they believe, sought and wanted because it is rare, valuable, or unique, but because the other, another one, someone else, seeks and wants it. Its value is in the eyes of the beholder, but that value is gained through the eyes of another.

Unfortunately this is what we do not know, what one never knows of himself, for we imitate so quickly, by a movement that is so spontaneous and primordial that it precedes all representation, all knowledge and consciousness of the fact that we imitate. Therefore what we first discover is conflict, the stubborn and senseless opposition of the other who suddenly, and without reason, seeks and longs for what we always already had sought. What we first discover is envy, jealousy and the unfair aggression of others, their invasion of our private domains. We fail to notice our own imitation and the movement by which we converge upon an object which another had designated to us as desirable. Desire itself is born of this imitation and of the beginning of the opposition that frustrates us, that prevents us from reaching the object we seek. Now given that, by hypothesis, this situation is rigorously true of all, in the beginning there can only be violence, strife, opposition, in the beginning there can only be chaos and disorder.

Order, argues Girard, arises out of that disorder through an exasperation and aggravation of the violence and conflict. Mimesis or mimetism is a strange thing. If it were strong enough to overrule or to overrun the instinctual fixations among a band of primates it would open to them a wide, nearly infinite, range of new behavior. All we need to suppose is that diverse individuals manifest various deviations in their behavior, deviations which through imitation can be extended to all members of the group and then be stabilized as new elements in their behaviorial repertoire. If this were the case it would also create for these primates an incredible problem of violence, for under this hypothesis mimesis would be sufficiently powerful to overrule the instinctual constraints that prohibit intraspecific murder. In those circumstances our band of primates will either disappear through mutual extermination, or, if they are to survive, some self-regulating mechanism must be found within the violence that threatens them.

Now as the mimetic rivals persevere in their opposition they progressively lose sight of the object that seemed at first to motivate their conflict, simply because the opponents, being active, quickly move to the front of the stage. The rivals, as their enmity grows, forget the reasons, the object, the cause from which the conflict, according to them, originated. Each one concentrates on his rival, and as they do this they become doubles of each other, each performing the same violent gestures as the other. When this happens, and as violence spreads through the whole community, an unexpected phenomenon may follow. If rivalry and mimesis to acquire one and the same object divides and pitches the members of the group one against the other, mimesis of the antagonist, the imitation of violence can unify them as all converge against one and the same enemy. Nothing, or only an insignificant difference, may predispose this one rather than that one to become the unique victim of the collective fury. But once this choice has been made and the victim has fallen, the most fantastic difference will be attributed to him. Why? Because the victim's demise will bring back peace to the community. Given that all the members of the community united, unanimous minus one, will vent and satisfy their rage simultaneously against the victim, they will effectively destroy what is for them the only cause of violence. As they do, peace will return. The victim will then appear to them as the unique cause of both the violence that preceded his death and the reconciliation that follows it. It will appear as the ruler and master of violence, as the source of all peace, of all life, and as the cause of all war, of all death. It will appear, in short, as sacred. Such is the origin of the social order according to Girard, the fountainhead from which the social order springs.

Here is a grand theory, one that has the dimension of the myths it deconstructs. But there is more to it than the epic greatness of origins, it also claims to be science, knowledge, an hypothesis. The theory postulates a self-regulating mechanism of violence, and the social order, it says, emerges from that self-regulation of violence. Violence and the origin of social order, paradoxically violence here is the origin of the social order and it is also what threatens to destroy it. Two questions remain to be answered: how does the order emerge? Why is the order that emerges a *social* order?

It is not enough that peace should return and that the calm be reestablished when the victim is destroyed in order for us to be able to say that an order has emerged or exists. That peace and calm must also be maintained and some structure or constraints must be imposed on the behavior of the agents. How will this miracle take place? The resolution of the mimetic crisis, the victimage that puts an end to it, rests on a misunderstanding, misrepresentation, or misrecogniton on the part of the agents who bring it about. First the agents do not see, or perceive, the mimetic origin of their conflicts, nor do they recognize their mimetic, quasi-automatic, dispersion, contagion, evolution to the scapegoating finale. Then the agents do not see the victim for what it is, the helpless victim of their own violence. Rather they see the victim as the lord and master of all violence, as the cause of violence at first, when they exterminate him, and then as the giver of the peace that follows. The whole crisis appears to the agents as having taken place under the sign of the omnipotent victim. That is why the victim is sacred. The social order springs from this misunderstanding. A misunderstanding that is in no way accidental, but that is, to the contrary, a necessary condition for the resolution of the crisis.

The agents, convinced of the victim's omnipotence, convinced that it has orchestrated the whole crisis, will reflect upon its evolution in order to understand and discover the cause of the violence that befell them, and the origin of the peace they now enjoy. It seems normal to assume that, in an effort to maintain the peace, they will prohibit certain gestures and behaviors which they associate with the evolution and aggravation of the violent crisis. This according to Girard is the origin of prohibitions in traditional societies. Simultaneously, in an effort to reap anew the benefits associated with the conclusion of the crisis they will reenact the gestures that hastened this conclusion and especially the immolation of the victim. This is the origin of rituals and of sacrifices. This singular genesis of rituals and prohibitions from the mimetic crisis would also explain the strange contradiction between them which anthropologists have often noticed: the rituals often prescribe exactly that which prohibitions prohibit. From the point of view which is ours this is not surprising, for the gestures which are associated with the evolution and aggravation of the crisis must be the same gestures of violence and conflict that brought about its resolution. Finally myths, according to Girard, would tell the tale of the crisis and of its resolution from the point of view of those who killed the victim.

Why, and in what way, is this emerging order a *social* order? It is, I believe, in two ways. First, it is not a natural order, because the restrictions, prescriptions, constraints, regularities which are imposed upon the agents' behavior do not result directly from their instinctual constitution but from the accidents which orient the evolution of the mimetic crisis. The second reason is that the evolution of the crisis and its resolution are predicated on the representations, misrepresentations, misrecognition which the agents have of the very process in which they are engaged. I have argued elsewhere that it is a necessary and sufficient condition to define a social system, order or interaction to say that it is a system, order or interaction which in its functioning rests on the representations which the agents, or elements, of the system or interaction have of the very interaction which they realize.[3]

This definition and characterization of the social nature of the crisis and of the order that proceeds from its resolution may seem, at first sight, to contradict some of Professor Girard's claim and assertions. In *Things Hidden Since the Foundation of the World*, Girard (1988) argues that the mechanism of victimage that puts an end to the crisis is at the origin of symbolicity (*le symbolique*), of symbolic language, in which one thing is given for another, represents another, the first symbol being the victim. The difficulty is that, at least in the continental philosophical tradition, i.e. Derrida, no separation is possible between the appearance of symbolic language and of representation. It follows, under this reading, that the evolution and resolution of the mimetic crisis would rest on what precisely is supposed to emerge from them: representation. The difficulty, I think, is only apparent. It stems from a confusion between two interrelated but different notions of representation. On the one hand representation, inasmuch as it is indispensable to the unfolding and conclusion of the crisis, is used in a way closely related to the way cognitive science uses that term. In this sense, representations are characteristic features of any type of cognitive system, be it human, animal or machine. On the other hand inasmuch as representation, symbolic language, is said to emerge from the process of victimage, what emerges is not Fodor's language of thought (Fodor, 1979) but the natural language which ultimately allows us to formulate theories about the language of thought.

Of course Derrida, clearly, would refuse to distinguish these two meanings of representation. I suspect, in fact, that he could use my interpretation of Girard to show how in Girard, as in Rousseau, the origin must always presuppose itself, to show that there is no origin, no starting point, to show, contrary to what Professor McKenna has claimed, that Girard's originary hypothesis does not escape his criticism of origin.

Be that as it may, what is, exactly, Derrida's critique or origin? How does it relates to the notions of sign, of representation, of voice, of writing, of history? For Derrida the idea of origin is part of a network of concepts, like eidos, ousia, sign, representation, history, substance, which all belong to Western metaphysics, or to onto-theology. Western metaphysics is essentially, or by definition, according to Derrida, the metaphysics of presence. That is to say, the metaphysics of being as being present. It is, in a sense, a nostalgia of pure presence.

Take as an example the concept of sign as defined by Saussure. A sign signifies as the signifier of a signified, whether it be a sense or a reference. A sign only signifies inasmuch as it relates a signifier to a signified, where the signifier stands for, takes the place of, is present instead of the signified. The gist of Derrida's critique is that ultimately there is no signified that is not itself a signifier. Why? For a relatively simple reason which is discovered through an analysis of Husserl (Husserl, 1962). What could be a pure signified that would not be, in anyway, a signifier, a sign of anything else? Evidently it could only be the thing itself, ideal or material, in flesh and blood so to speak, the thing in its pure presence. Now such a thing, argues Derrida, cannot be the object of any possible experience, or if it can, that experience cannot by any means be communicated. It is exterior to the domain of signification and hence, by definition, cannot be the referent or sense, the signified of any signifier. Why? Because this object would be the object of a perfectly singular experience, in order to be communicable, to gain a signification, it must become an ideal object, an eidos, that is to say, a repeatable entity that can be associated to the signifier in all of its use, and which is in the best of cases, which is truth, associated to the signifier every time it is used. But clearly this cannot be the thing itself, the absolutely singular experience of its pure presence, it can only be a sign of that presence, a signifier that refers to it. Conclusion: such an experience cannot signify anything, it is, by definition, without the domain of signification. Conclusion of this conclusion: there is no signified that is not already a signifier. There is no exit from the infinite interplay of signifiers each one referring to another. It follows that we are enclosed within a system of signs, or if you prefer, of signifiers, and that there is in a way a closure of this system. It follows finally that truth as adequation between what is in the mind and what is in the world is an illusion, the illusion par excellence of the Occident, the illusion of a pure presence, God, Nature, Man, that founds and guarantees the system of signs, the illusion of an origin that transcends, orders and guards the domain of signification. Deconstruction as a philosophical attitude is at war against this illusion in an effort to restore an infinite play of signification which, by definition, was never lost, but that was perhaps, in some obscure way that remains to be understood, dissimulated by the illusion of a pure origin.[4]

My point is not to discuss the advantages or defects of such a philosophical attitude but rather to draw your attention upon two things. The first one is that deconstruction as a strategy, and deconstruction according to Derrida is a philosophical strategy, repeats the characteristic gesture of Rousseau and Husserl concerning the question of origin, which is to discard all facts. Hence my first claim that Professor McKenna's relative silence about the origin of social order is not an accident, but the necessary consequence of his endeavor to reconcile Girard and Derrida. The price to pay for this reconciliation is to overlook the fact that Girard makes a precise hypothesis concerning the origin of social order, an hypothesis which is to be judged by its fertility and simplicity to organize data, and ultimately by its capacity to withstand empirical tests. This price as we will soon see is not without consequences.

Towards the end of this paper Professor McKenna suggests an oecumenical comparison between Girard's conception of the role of violence at the origin of the social order and Derrida's reading of the nuclear holocaust in 'No Apocalypse, not now (Full speed ahead, seven missiles, seven minutes)'.[5] The comparison rests on the fact that for Derrida the nuclear catastrophe offers us the "trace of what is entirely other, *trace du tout autre*" (p. 72) and that this "wholly other" though it is not named as the sacred by Derrida, "he doesn't have to, for that is what all the philosophers since Hegel have named it" (p. 73). And since for Derrida this absolute trace is the "only ⟨⟨subject⟩⟩ of all possible literature, of all possible criticism"[6] it holds the same place as violence for Girard which is "the true subject, the active and passive agent, the agendum and dilendum of all cultural institutions, of institutional formation and dissolution" (p. 73) and violence, as we know, for Girard is the sacred.

Unfortunately the comparison does not hold, and this for at least two reasons. The first one simply is that total remainderless destruction cannot, for very evident reasons, be the agent of all cultural institutions, of all institutional formation and dissolution. Or if it can, it can only be as a trace, in its absence. As Derrida said, "the nuclear age is ⟨⟨*faboulously textual* through and through⟩⟩" (p. 72) and it is textual because its referent is by definition absent; which is also why its referent is the "only ultimate and a-symbolic referent, unsymbolizable, even unsignificable; that is . . . the remainderless and a-symbolic destruction of literature".[7] This brings us to the second reason why the comparison is not successful. For Girard the threat of the nuclear catastrophe hanging over our heads is not the trace of what is "wholly other", rather it is the clear revelation that what we always conceived as the wholly other, the sacred, absolute violence, is in fact nothing but our own violence projected outside of ourselves and dehumanized. In that sense, from a Girardian point of view, Derrida's attitude once again repeats the characteristic gesture of sacrificial thinking, which is to dehumanize human violence, to transform it into an "unassimilable wholly other".[8]

The second point on which I would like to draw your attention is slightly different though it is also related to Derrida's conception of origin. My claim is that implicit in Derrida's text is a concept of origin that cannot accomodate his own criticism of origin. A few quotations will be necessary in order, if not to show it, at least to indicate why one may be tempted to think that such is the case.

> Logo-phonocentrism is not a philosophical or historical mistake in which accidently, pathologically would have plunged the history of occidental philosophy, or even world philosophy, but rather it is a necessary and necessarily finite movement and structure: the history of the possibility of *symbolicity* (*le symbolique*) in general (before the distinction between human and animal and even the living and non-living); history of *différance*, history as *différance* (1967a, p. 214).[9]

> The energy of occidental theatre can thus be captured in its possibility, a possibility that is not accidental, that is for the whole history of the Occident a constitutive center, a structuring locus (1967a, p. 365).

> This concept (the vulgar concept of time) that determines all of classical ontology, was not born of a philosophical error or theoretical failure. It is interior to the totality of the history of the Occident, to what unites its metaphysics to its technology (1967b, p. 105).

It is this history (the history of the presence of life to itself) (as period: not as a period of history, but of history as period) that closes itself at the same time as the form of being-to-the-world which we call knowledge. The concept of history is thus the concept of philosophy and of *épistémè* . . . . What exceeds this closure *is nothing*: neither presence nor being, neither history nor philosophy; but something else which has no name, which beckons in the thought of this closure and conducts here our writing. Writing in which philosophy is inscribed in a place and a text which does not command (1967b, p. 405).

Because it has always already begun representation has no end. But we can think the closure of what has no need. The closure is the circular limit within which the repetition of difference repeats itself indefinitely. It is the space of its play (1967a, p. 367).

Only two more and I promise you a little rest.

We could show that the concept of *épistémè* has always called the concept of *istoria* if history is always the unity of a becoming, as a tradition of truth in the presence and presence to one's self, towards knowledge and self-consciousness (1967a, p. 425).

The event of the rupture, the disruption to which I was alluding at the beginning, would perhaps have happened when the structurality of the structure began to be thought, that is, repeated, and that is why I said that this disruption was repetition, in every sense of the term (1967a, p. 410).

These two last quotations are interesting inasmuch as the first one (1967a, p. 425) is a criticism of the notions of history and épistemè, while the second offers praise for a certain event, the irruption of structuralism and post-structuralism into philosophy and social sciences. The difficulty, of course, is that the reader is somewhat at loss to distinguish what is being praised from what is being criticized. After all what does it mean, especially in Derridian lore, to say that "the structurality of the structure began to be thought, that is, repeated", if not that it became present to itself, conscious of itself, knowledge? What else can it mean in view of the preceding pages of the quoted text which inform us that the "structurality of the structure though it has always been at work, was ever neutralized and reduced; by a gesture which consisted in giving it a center, in relating it to a point of presence, a fixed origin" (1967a, p. 409). We can recognize here the illusion of a pure presence that dissimulated the infinite play of signification. Yet by what means can we recognize an advantage to the end of this dissimulation if not by the very schema that justifies the condemnation of history and knowledge?

It can be objected that my criticism is somewhat unfair. Derrida has always said that deconstruction is caught in a circle which is due to the fact that, to criticize metaphysics, it only has at its disposal the concepts of metaphysics (1967a, p. 412). Hence we should not be surprised to find in deconstructivist texts the type of contradictions I have maliciously uncovered in Derrida's own work.[10] Yet the criticism, though unfair, may not be entirely irrelevant. Why? Because we have reasons, both textual (1967a, p. 411) and conceptual, to believe that this sudden revelation of the structurality of the structure is not accidental. In fact, as should be clear by now, the point of the first five quotations was to show that the infinite interplay of significations was, in some ill-defined way, placed under the sign of a certain necessity. It is this idea of necessity, in its relation to the concept of closure, that I now want to question.

This necessity, as it is easy to verify by consulting many texts (1967a, p. 410; 1967b, pp. 288—289; 1972, p. 111), always takes the form of a repetition, is always thought under the concept of repetition, that is as "infinite substitutions in the closure of a finite set" (1967a, p. 423). It is these infinite substitutions in a finite set that interest me. They point to an idea of origin, whether it be event, structure, or presence, in which everything is always already given, in which history as the unfolding of the structure, or as the infinite interplay of significations is always already contained. They point to an original donation from which all necessarily proceeds or recedes and which in its closure excludes that anything *new* should appear. Yet is not that original donation the very concept of origin which Derrida criticizes? Is it not that exclusion of what is new that defines history as the reappropriation of an original presence and knowledge as the repetition of what is already here? Or is this criticism unfair? It is only an hypothesis, let us verify it upon another text.

At the beginning of the third Chapter of *De la Grammatologie*, entitled 'Of Grammatology as a positive science', Derrida asks: under what conditions is a grammatology, a science of writing, possible? He answers, under the condition of knowing what is writing, where and when it began. Where and when are questions of origin and these can be understood in two ways, as empirical questions and as the question of the essence. And the question of the essence, adds Derrida, must be answered, for it is its answer which guides the empirical research.

"We must know *what is* writing in order to be able to ask, while knowing what we are talking about and *what is* in question, where and when writing begins ? What is writing? By what can we recognize it?" (1967b, p. 110). But is this really necessary? Is there not another form of questioning, linked to a different conception of origin, that avoids the last two questions, "What is writing?", "By what can it be recognized?". In other words, is it true, as Derrida pretends, that "the question of origin is first of all indistinguishable from the question of the essence"? (*Ibid.*) For the question of origin, "when and where" can also be formulated in a different way, "Whence came writing", "From where came writing?".

This question does not presuppose the question "What is writing?", the "onto-phenomenological question in the rigorous sense of the term" (*Ibid.*) because the question simply is: How came to existence what we commonly call writing? In order to answer this question it is not necessary to know before hand, in the full sense of determining its essence, what is writing. For the question does not imply that writing is, in any sense of the terms, anything simple and unitary. Rather the question asks, through what process did what we commonly recognize as writing, including perhaps Derrida's 'arch-writing', come into existence? The question does not presuppose that a unique essence of writing is present throughout this process. All it presupposes is a set of elements over which are distributed numerous traits of resemblance that allow us to trace a filiation from what we would perhaps not recognize as writing to what we actually name writing. In other words what we call writing would be defined by family resemblances and different forms of writing would form a polythetic class. In that very particular sense it is the question of the origin that would answer the question of the essence and not the question of the essence that would determine the question of origin.

Now what conception of origin is implied here? Clearly what is ruled out is the idea of an original donation that always already pre-determines the evolution, unfolding, play, of logo-phonocentrism, of Western metaphysics, in the infinite yet bounded space of its closure. What is implied is an evolutionary process of some sort, a mechanism of complexification, which allows the emergence of new properties which are not present, or in anyway contained, in the properties of the elements which constitute the system, that is to say, at the origin. Such an inquiry can only be retrospective and factual, historical in other terms, for it is impossible to determine beforehand the future evolution of a system which is marked by the emergence of radically new properties.[11] In these circumstances, knowledge of the essences becomes a very secondary and perhaps misleading matter.

During these last few years many of us have argued that Girard's theory is precisely of that type, a morphogenetic theory which allows the modelization of new phenomena which were not actually contained in the starting point.[12] In fact this should already be clear from my presentation of this theory in the first part of this commentary.

Once again it may be argued that my criticism is unfair, that Derrida knows very well, better than anyone else, that writing and *différance* have no essence. This is certainly true. Yet my point is not that Derrida does not know this, but that he falsely concludes from this that the question of origin is impossible. For he thinks that the question of origin necessarily presupposes the question of the essence. Simultaneously I have tried to show that his concept of closure presupposes the idea of an original donation which he otherwise criticizes and rejects. Hence my second claim, that implicit in Derrida is a concept of origin which cannot accomodate his own criticism of origin. Perhaps this criticism is unfair. Or are these contradictions the price to pay for the choice, not the inescapable necessity, of criticizing metaphysics with the concepts of metaphysics only? "We must begin by discarding all the facts."[13]

*University of Waterloo, Ontario*

NOTES

[1] From now on all references to Professor McKenna's text, 'Supplement to Apocalypse: Girard and Derrida' will be given between parentheses.

[2] Aristotle, *Poetics*, 4.

[3] Cf. my 'Chickens, Cyclists and Scientists', forthcoming.

[4] It is evident that this infinite play of signification could never have been lost, hence the problem of under which category, if not that of knowledge, in the classical sense (according to Derrida) of the repetition and recovery of the origin, can we think the value of the deconstructivist's awareness of this infinite play? This problem will reappear later on.

[5] *Diacritics*, XIV, 2, 1984.

[6] *Ibid.*

[7] *Ibid.*

[8] *Ibid.*

[9] Page numbers refer to Derrida, 1967a, 1967b or 1972.

[10] Cf. Derrida, 1967a, p. 412.

[11] On these problems and the possibility of predictive theories in such a situation see Popper, *The Open Universe*.
[12] See Dumouchel, 1983; Dupuy, 1983; Livingston, 1988.
[13] Rousseau, *Discours sur l'origine de l'inégalité*, quoted by McKenna (p. 45).

## REFERENCES

Aristotle: *Poetics*.

Derrida, J.: 1967a, *L'Ecriture et la Différence*. Paris: Seuil.

Derrida, J.: 1967b, *De la Grammatologie*. Paris: Minuit.

Derrida, J.: 1972, *La Dissémination*. Paris: Seuil.

Derrida, J.: 1984, 'No Apocalypse, Not Now (Full speed ahead, seven missile, seven minutes', *Diacritics*, XIV, 2.

Dumouchel, P.: 1983, 'Mimétisme et Autonomie' in *L'Auto-organisation, de la Physique au Politique*. Paris: Seuil.

Dumouchel, P.: forthcoming, 'Chickens, Cyclists and Scientists'.

Dupuy, J.-P.: 1982, 'Mimesis et Morphogenèse' in *Ordres et Désordres*, Paris: Seuil.

Fodor, J.: 1979, *The Language of Thought*. Harvard University Press.

Girard, R.: 1988, *Things Hidden Since the Foundation of the World*. Athlone.

Husserl, E.: 1962, *L'Origine de la Géométrie*. PUF. Introduction et traduction de Jacques Derrida.

Livingston, Paisley: 1988, 'Durkheim et Girard' in *Violence and Truth*. Stanford: Athlone.

Popper, K. R.: 1982, *The Open Universe*, Rowman and Littlefield.

PAISLEY LIVINGSTON

# GIRARD AND THE ORIGIN OF CULTURE

## 1. GIRARD AND DECONSTRUCTION: A RESPONSE TO McKENNA

I certainly share Andrew McKenna's enthusiasm for some of René Girard's ideas, and admire his effort to mediate between Girard and aspects of post-structuralist thought. Yet I am afraid we disagree on a number of very basic points. This disagreement does not concern the tenets of some kind of Girardian orthodoxy; rather, what is at stake are different ways Girard's insights may be interpreted and extended. In this regard, the most basic difference between McKenna's position and my own involves my skepticism about the proposed bridge between deconstruction and Girard's anthropology.

McKenna describes the relations between essentially two positions, while I believe that there are other important positions that must be taken into account if we want a more comprehensive view of the options available to us. McKenna perceives some fundamental similarities between the theses of Girard and aspects of texts by Derrida, and argues that these two thinkers are alike in rejecting metaphysical and mythical stories of origins. Thus a key line in McKenna's paper is his statement that "Girard's originary scenario can accomodate a post-structuralist critique of origins because it posits that origin and differance are one, that representation is the by-product of an originary differance, of mimetic desire, rather than the representation of an originary presence, of an origin of any kind" (p. 55). I read this as a way of positioning Girard and Derrida on one side, and metaphysical concepts of origins on the other. Yet the intellectual landscape looks very different when we take into account another major set of positions, which may be referred to rather loosely as the naturalistic or natural scientific ones. Some of these positions are the infamous sociobiological and reductive evolutionary accounts of the emergence of human culture, and in relation to that sort of science fiction a 'post-structuralist critique of origins' is no doubt a good thing (although one may also consider that the most rigorous and compelling critiques of



*Francisco J. Varela and Jean-Pierre Dupuy (eds), Understanding Origins, 91–110.*

sociobiology have been those provided by natural scientists capable of showing what *does not* follow from evolutionary theory). In any case, it is crucial to note that a natural scientific approach to the emergence of humankind does not entail any theory in which a putative survival of the fittest is the single mechanism.

In short, I think the differences between Girard's anthropological project and the various post-structuralist deconstructionisms are far more important than any similarities one may find. Moreover, it is overly limiting to suggest that we must choose between a post-structuralist blend of Girard and Derrida, on the one hand, and an unspecified 'metaphysics' on the other. Many other salient approaches and positions are simply left out in such a scheme of things, most notably, a range of non-reductionist natural scientific perspectives.

## 2. A NATURALIST FRAMEWORK

In his chapter on hominization in *Des choses cachées depuis la fondation du monde*, Girard certainly appears to adopt a naturalistic framework for his hypotheses about the emergence of human culture. He explicitly claims that he wants to resolve the dispute between ethologists and ethnographers, with their respective emphases on the continuity and discontinuity between the animal and human orders. He writes that ethologists are correct in "protesting against the extraordinary insularity of culturalist and structuralist ethnography, particularly its absolute refusal to view human culture as part of nature, as well as its properly metaphysical conception of symbolicity" (1978a, pp. 100—101). He refers to the activity of 'instinctual mechanisms' and speaks of the increasing size of the hominid brain, while calling more generally for an anthropology that would be *enracinée dans l'animal* ('rooted in the animal'). At the same time, Girard castigates those for whom 'Dame Evolution' serves as a quasi-magical device for explaining culture (1978a, pp. 97—98). This is an important insight because it is precisely the rigid and reductive assumptions about evolutionary process that ruin much of the naturalistic theorizing about the emergence of homo sapiens (for examples, see Fox, 1975; Lockhard, 1980; Maxwell, 1984; Wilson, 1980). Many of these evolutionary theories are 'just-so' stories that find no compelling evidence in what physical anthropology presently can reveal about the nature of the Primate Order and its environment around a couple of million years ago.

What, then, is Girard's alternative to the false choice between a reductive evolutionary account and the idealism and particularisms of culturalist ethnography? Girard describes a hypothetical natural starting point for culture in a scenario that has multiple presuppositions. One of the most basic premises in his discussion of hominization is that language, symbolism, and representation were not always already there, but emerged as a result of natural processes. He posits an 'origin' only if we mean by this an account of the possible antecedent conditions of the emergence of a particular set of phenomena. These conditions are extremely complex, and reference to them in Girard's hypothesis entails a committment to many far-reaching assumptions. Thus if Girard speaks of 'originary events', it is important to be aware of what has to be in place if these events are going to be able to play the role described in the scenario. Most notably, some very complex natural conditions and behavioral patterns must be in place for the scenario to be realized, and these are conditions that could only make sense within a naturalist account.

The naturalistic framework in question here is a matter of a particular background of problems and assumptions. It also involves a number of epistemological commitments. The latter specify, for example, that Girard's anthropology should be approached as a scientific project in which hypotheses are formed so that they may be evaluated in terms of available evidence. In keeping with such an approach, Girard has indeed employed the language of hypothesis and verification; he frequently castigates areferential views of language as well as skeptical and irrationalist epistemologies. His work, then, need not be approached as an exercise in speculative metaphysics. This means that we will not try to decide, as a matter of principle, whether science can one day successfully explain the historical emergence of human thought. Given the paucity of the record and the extreme complexity of the phenomena in question, there is no reason to have great confidence in any of the present conjectures about hominization and the historical genesis of symbolicity. Nor, however, is there any reason to assume that these phenomena are unknowable in principle because somehow veiled by the mysteries of *Nachträglichkeit*. Thus in what follows, I shall be discussing Girard's analyses uniquely within a very general naturalist framework that assumes a realist scientific epistemology (for details see Boyd, 1973, 1979, 1983, 1985a, and 1985b; Livingston, 1988a). I will not be taking up the theological elements in Girard's philosophy of history (for a critique of the latter, see Livingston, 1988b and forthcoming).

## 3. THE GENETIC HYPOTHESIS

In order to provide a first, schematic presentation, we may observe that the following moments are essential to Girard's basic originary scenario (see his 1972, 1978a, 1982, 1985):

(a) The natural emergence or evolution of primates. Their social order is regulated by instinctual behavioral patterns, and above all else, by patterns of dominance and submission (this stage presupposes the existence of a natural environment, the emergence of life, and so on).
(b) The natural emergence of hominids characterized by increasing degrees of mimesis, and in particular, by so-called 'appropriative mimesis'. This mimesis generates typical sequences of interaction involving rivalry and violence.
(c) As a result of this emergence of intraspecific rivalry among hominids, there occur crises of violence that are disruptive of activities and relations within the hominid groups.
(d) The resolution of these crises of violence occurs by means of the mimetically-driven convergence of the group's behavior on the interactive patterns of victimization: the configuration of 'the many and the one' is the basic pattern of social order, realized through this mimetic convergence.
(e) As a consequence of (d), there emerges a form of 'non-instinctual attention', beginning with mythical attitudes toward victims of collective violence (distorted images of the victim are wrongly perceived as cause of the crisis and of its resolution — whence the structuring of sacred representations around the two poles of maleficence and beneficence). The symbols of the sacred are in turn the foundation of the diversity of cultural institutions, starting with myth, ritual, and sacred prohibitions.

Girard argues that his concept of mimesis makes it possible to raise the question of hominization and the emergence of symbolicity within a naturalistic framework. An understanding of how this mimesis works is crucial to the whole theory, for this mimesis, while it has a natural origin, is in turn the origin or determining condition behind the patterns of action and representation held to be constitutive of a specifically human social order. In Girard's theory as I understand it — and as I would prefer to see it developed — temporal and etiological priority belongs to animal mimesis, not to desire, representation, victims, the sacred, violence, and so on, because it is this mimesis that is posited as a necessary causal condition of phenomena associated with all of these other terms. And this mimesis itself has an origin insofar as it is said to have evolved or emerged as a natural reality within the natural universe.

The basic Girardian scenario of the emergence of the human socio-cultural order represents a bold schema that avoids the glaring weaknesses of many alternative speculations. Yet much more work remains to be done before it can be presented as a detailed and comprehensive hypothesis. Four main lines of inquiry, each of which represents a major research project, can be identified at this time:

(i) The scenario has not been related in any detail at all to the voluminous literature in physical anthropology dealing with the fossil record. Unlike many natural-historical accounts, the Girardian scenario has not been correlated, even at the level of preliminary speculation, to what has been conjectured about the different periods and environmental events of the prehistoric past. Girard states, for example, that meat eating, war, cannibalism, and hunting developed early on in the process of hominization (1978a, p. 94). These claims may be accurate, but they have been intelligently contested (e.g. Potts, 1984) and more research is required. (It should be noted that the documentation on hominization in Girard, 1978a is very sparse; he relies on two articles in Montagu, 1962 which hardly represent the state of the art on the question; Girard cites (p. 95) a text by Washburn (p. 17) that posits without argument a "drive for first place" among the early hominids. Such a statement recalls the reductive claims of sociobiology. For rather different views on the place of drive constructs in contemporary motivational theory, see Madsen, 1974; Satinoff and Teitelbaum, 1983; Toates and Halliday, 1980; Toates, 1986.)

(ii) Although statements about primate social orders figure in Girard's speculation, these claims have not been supported by extensive reference to ongoing work in ethology. Example: bold assertions about stable dominance/submission patterns among apes; Girard, 1978a, pp. 99—100; compare Snowdon, 1983, pp. 82—83, but also Hinde, 1983. Systematic reference to the literature on primate behavior and social organization is necessary if such claims are to be supported or revised.

(iii) The theory's relation to today's various evolutionary theories also needs to be specified. An explanatory model for the hypothetical natural emergence or evolution of heightened mimesis (stage (b) above) must be provided. In other words, if Girard is right in his criticisms of the use of Dame Evolution in various sociobiological theories, it nonetheless remains the case that some kind of evolutionary processes must play a major role in his scenario, and their nature needs to be specified (see, for example, Sober, 1983).

(iv) Finally, the theory of hominid agency requires further specification so that a more detailed account of stages (b) through (e) above can be provided — even at the level of a speculative construction of hypotheses. Several major questions surround this theory of agency. Most generally, we need to say in more detail how the primitive kind of mimesis is supposed to work. To that end, it is necessary to confront the following interrelated questions: What is the relation behind hominid mimesis and the higher cognitive capacities that emerge at the end of the process of hominization, and more specifically, how is this primitive form of mimesis related to representations or beliefs? What is this motivational system's relation to the type of action that is motivated or guided by thoughts? How does the primitive form of mimesis generate conflict as well as a particular manner of resolving conflict in the group? What is its relation, finally, to the form of mimesis that Girard designates as mimetic desire?

My focus in what follows will be on the cluster of issues just mentioned under rubric (iv). For a detailed evaluation and reconstruction of Girard's views of the *post*-cultural phenomena associated with mimetic desire, see Livingston (forthcoming).

## 4. CONSTRAINTS ON THE GENERATIVE SCENARIO AND ON THE THEORY OF HOMINID AGENCY

In thinking about various possible interpretations of, and alternatives to, Girard's generative scenario, it is important to have clearly in mind certain features of the overall argument. Girard reasons backwards from a particular set of assumptions about the invariant nature of the first human cultures to a hypothetical scenario for their genesis. The argument takes the following form: if all (known) archaic cultures had the features, *x, y, & z*, then any acceptable hypothesis about the genesis of culture must posit conditions capable of giving rise to (at least) *x, y, & z*, and *not* to anything incompatible with *x, y, & z*. We then look to the available evidence about the hominids, primate behavior, and so on to try to hypothesize a set of conditions satisfying this requirement.

The assumptions about *x, y, & z* amount to a particular account of the invariant features of the institutions of known archaic cultures, these features being inferred from documentation pertaining to mythology and ritual practices. There is much to be said about the strengths of this account and about potential objections to it; it is obvious that if we do not accept Girard's assumptions about the invariant features that define 'primitive culture', that is, if we think that *x, y, z* are inaccurate and need to be replaced by *q, r*, and *t* (or abandoned entirely), then the whole discussion shifts. This is a vast and crucial topic, but it will not be my present focus. What is more important in the context of a discussion of mimesis is a rather different question: what are the constraints that the Girardian theory of archaic cultural formations weighs upon the details of particular scenarios for the originary sequence? Clearly, these constraints are determined by the nature of the effects to which the originary scenario is supposed to lead, and therefore it is important to have *x, y, & z* clearly in view. Any scenario which cannot plausibly produce these particular effects is to be ruled out in advance. This is of course only a first condition, and should it be satisfied by several different scenarios (which is most likely), it would then be necessary to discuss the criteria in terms of which they could be further evaluated.

Essentially, in Girard's theory, the effects that any 'properly originary' hypothesis must plausibly generate are as follows. The dispositions of the hominids must be described in such a manner that (1) their behavior will tend to give rise to the phenomenon of collective murder; (2) these murders must occur under conditions such that they give rise to the emergence of 'sacred significations' as described by Girard, these significations being the necessary (but not sufficient) psychological bases of the central cultural institutions (ritual and taboo, kinship and kingship, etc.). (The other bases of culture involve, quite importantly, the mimetic dispositions and motivational systems of the individuals.) In a nutshell, the sacred significations in question hinge upon the group's belief in the transcendence or alterity of supernatural entities, the images of which have their real basis in the victim. Moreover, there is the related polarization of sacred beliefs — involving distinctions between maleficent and beneficent entities, powers, and practices.

Girard typically implies that the members of the group must in some sense remember the violent crisis. They dread its recurrence and have a positive recollection of the moment of well-being experienced at its end. At the same time, the group's memory of these events is distorted, and thus the prototypical sacred belief involves a fundamental *méconnaissance* or misrepresentation of the real basis of these beliefs. The collective murder is not recalled as such; in the group's representations, the real victim is replaced by a maleficent agent capable of being the sole cause of the terrible crisis of violence, the agent whose demise or departure brings about a magical transition to a state of order. It would seem that in many instances, the perception of the single being who was the victim is replaced, in sacred thought, by two entities: the one is a malevolent and terrible source of chaos, while the other is the heroic enemy and slayer of the monster. Oddly enough, then, the first essential non-animal effect, the *x* in the *x, y, & z*, involves a *fundamental misrepresentation* of the identity of a hominid cadaver. In other words, following Girard's account, a necessary condition of the emergence of properly human culture is a particular kind of false belief, a primary illusion, one that is religious in the Durkheimian sense.

The foregoing remarks give us a clearer sense, then, of what Girard's theory of hominid mimesis must plausibly produce if it is to play the role allotted to it in his genetic hypothesis. Roughly put, the hominid mimesis in question must generate (a) certain patterns of interaction; and (b) a disposition for certain types of false beliefs about those patterns. This is clearly a tall order, and in what follows, I will be focusing on some of the difficulties that surround this scheme. I will not be trying to propose any conclusive arguments for or against the viability of Girard's scenario; rather, I shall strive to clarify some of the issues and point out what I see as some of the stronger and weaker approaches to them. More specifically, it will be argued that appropriative mimesis alone is unlikely to fulfill the task that Girard assigns to it and that other stipulations need to be provided.

## 5. THE NATURE OF PRIMARY MIMESIS

A form of mimesis, then, is hypothesized as being the generative mechanism behind a vast range of behavioral and interactive patterns, personality formations, beliefs, attitudes, and symbolic forms. Indeed, Girard wants to say that it is this mimetic feature of hominid beings that produces those specifically human phenomena that may be associated with the emergence of culture (e.g. religiosity or the sacred). Mimesis is what produces the sequence of interactions that lead to the first non-instinctual attention (1978a, p. 109). It is this attention that signals the origin of a property human social order, that is, one based on the symbols of the sacred: "Mimetic phenomena provide the common ground between animal and human society as well as the first concrete means to differentiate the two, concrete in the sense that all observable analogies and differences between the two types of organization become intelligible" (1978b, p. 35).

What, then, is this mimesis? What does its operation presuppose? Many of Girard's characterizations of mimesis are negative, for he wishes to dissociate his model from the various earlier notions of imitation with which it may be confused. A first point that should be underscored is Girard's insistence that the truly originary form of mimesis should not be referred to as a form of desire. It may be that mimetic desire becomes prevalent at a later stage in human history, but it is not already there at the origin: Girard points out that at the "elementary level the word desire is not appropriate" (1978b, p. 32; see also 1978a, pp. 308—309). But why insist on a distinction between mimetic desire and the form of mimesis that Girard calls *la mimesis d'appropriation*? The response has to do with the role of this primitive mimesis in Girard's theory. While mimetic desire presupposes the existence of a socialized human agent living within an established culture, the functioning of appropriative mimesis does not rest upon this assumption. Thus it can play a crucial part in a non-circular hypothesis about the genesis of cultural phenomena. As a kind of mimesis that is truly 'rooted in the animal', Girard's notion of appropriative mimesis cannot be reduced to classical philosophical notions about teleological actions that are guided by a subject's representations. The agent of Girard's primitive mimesis is not a 'subject' of action in this sense. Girard thereby seeks to avoid the kind of circularity a deconstructionist expects to find in any theory of the origin of representation (see Lacoue-Labarthe, 1978, p. 18). The agent does not represent to itself the model's actions prior to the generation of its own mimetic response (otherwise, there would a social representation or belief serving as the condition of the origin of social interactions). Girard writes that "This mode of imitation operates with a quasi-osmotic immediacy necessarily betrayed and lost in all the dualities of the modern problematics of desire, including the conscious and the unconscious. The desire for the other's desire has nothing to do with the Hegelian desire for recognition" (1978c, p. 89). In another relevant passage, Girard writes that "all subsequent mimetic effects are ultimately brought back to the radical simplicity of the primordial mimetic interference. When any gesture of appropriation is imitated it simply means that two hands will reach for the same object simultaneously: conflict cannot fail to result" (1978b, p. 32). Girard claims that human beings have no monopoly on mimetic rivalry, but he adds that in the human species, this kind of conflict is no longer restrained by an instinctual inhibition of intra-specific murder. Thus it can generate a 'fight to the finish': "an increased mimetic *drive* corresponding to the enlarged human brain must escalate mimetic rivalry beyond the point of no return" (1978b, pp. 32—33, my emphasis). In the end, the "mechanism of symbolicity must be triggered by the murderous exasperation of mimetic rivalry" (1978b, p. 33). In the present context, the following passage is crucial, and must be quoted at length:

> It is Plato who determined the cultural problematic of imitation once and for all, and this problematic is a mutilated one, lacking an essential dimension: the acquisitive dimension, which is also the conflictual one. If the behavior of certain superior mammals, apes in particular, seems to prefigure that of mankind, this is perhaps almost exclusively because of the large role already played in these species — but not a role as large as in man — by appropriative mimesis. If an individual sees one of its congenerics reach out its hand toward an object, it is instantly tempted to imitate this gesture . . . the animal is a kind of brother because he is shown to be subjected to the same fundamental servitude as humanity is — that of warding off the conflicts that can be provoked when two or more hands converge on the same object" (1978a, pp. 16—17).

By referring to an immediate and mechanical form of mimicry, Girard seeks to cut beneath the category of the 'subject' and what Derrideans may refer to as a metaphysics of re-presentation. Can such a strategy succeed? It cannot be denied that for there to be even the most minimal form of behavioral matching or 'monkey see, monkey do', the monkey must see before it can do. In other words, matching behavior cannot be explained in terms of some mysterious, pre-established harmony, and thus some form of perceptual linkage must be at work. But the necessity of such a linkage is only a valid objection if one believes that all instances of animal perception involve representation in some strong sense. In fact it is wrong to think that some properly human representational capacity must be at work in all forms of behavioral matching – the facts are on Girard's side in this regard (for documentation see Aronfreed, 1969). Yet one could object that the necessity of a perceptual linkage contradicts Girard's reference to the immediacy of the matching phenomenon. Clearly, this potential objection places a constraint on the way in which the notion of appropriative mimesis can be specified and employed if it is to retain its role in a theory that designates it as a truly primitive behavior. And a key aspect of this constraint, which Girard does not really address in detail, is the non-representational characteristic of the perceptual linkage between imitator and model.

## 6. GENERATING SACRED BELIEF

Our question, then, concerns Girard's admirable attempt to balance elements of continuity and discontinuity in his model. How can a primitive form of mechanical mimesis produce such prodigious cognitive effects as a false belief in the supernatural powers of a particular entity, who is no longer even recognized by its fellows as a member of the same species? This last question should be underscored, for it is important to recognize that what is said to emerge from the crisis of violence and its resolution is a set of typical beliefs and attitudes having the victim as their misrepresented object. These beliefs concern the victim's responsibilities and supernatural powers (as cause and cure of the crisis) and are the very foundation of the sacred in Girard's view – which implies that for Girard they are also the foundation of all properly human social orders. The first non-instinctual attention – if it is indeed the first – comes right on the heels of a crisis having purely mechanical determinations, those of appropriative mimesis; yet this crisis and its resolution are said to produce rather prodigious representational effects insofar as they lead to the group's misrepresentation of the victim, to its belief in his or her 'guilt' as well as to an amazed recognition (*reconnaissance émerveillée*) of the peace following the act of collective murder. To the extent that one stresses the non-representational and non-intentional aspects of the primordial mimesis, a rather mechanical theory of hominid agency is embraced; yet this analysis is conjoined to a description, in 'high' cognitive language, of the results of the scapegoat mechanism, results amounting to an ensemble of beliefs, attitudes, symbolic practices, and actions that are at once cultural and representational. And it must be the same creatures who experience the violence and the crisis caused by their own mimesis as well as the marvelous beliefs that surround the victim. We must remember that only in later situations, after the emergence of culture, can a belief in the victim's guilt actually motivate a unanimous turning of violence against one party: no such beliefs are at work in a unanimity generated by a non-representational mimesis. Beliefs in the victim's responsibility come later as a representational effect of this more mechanical form of mimesis.

A purely mechanical form of agency determines a behavioral sequence having highly representational and complex cognitive effects. There is no a priori reason why this is impossible, but one must still wonder whether there is a danger of overloading the bridge between the animal and the human that the theory is meant to erect. On the one hand, it could seem that creatures driven by fairly primitive drives and mimetic facilitations of these drives suddenly acquire higher cognitive capacities (beliefs, memory, etc.); on the other hand, it could seem that the first non-instinctual attention must have been preceded by some fairly complicated representational capacities. Girard's most lengthy explicit statement on this issue reads as follows:

> To the extent that the attention I am talking about is awakened, the victim is associated with the emotions stirred up by the crisis and its resolution. The striking experience crystallizes around this victim. As weak as it may be, the 'consciousness' that the participants have of the victim is structurally linked to the prodigious effects that accompany the victim's passage from life to death, as well as to the spectacular and liberating reversal that takes place at the same time. The only meanings that may appear are those of the double transference, that is, the meanings of the sacred, those that attribute to the victim active responsibility for the whole affair. But it is necessary to conceive of stages, perhaps the longest stages of all human history, in which these meanings are not yet really there. It is necessary to answer that we are always on the way to the sacred as soon as the call of the emissary victim is heard, no matter how feebly this may be, but there are not yet any concepts or representations.
>
> It is not necessary to think that the machine that awakens attention works immediately; we can imagine a large number of misses, or near misses [*des 'coups pour rien' ou presque rien*]. As rudimentary as these effects may be, these are what is required for a control of an excessive mimesis; it is sufficient to admit that these effects are as weakly cumulative as one wants to affirm that we are already on the way to the human forms of culture (1978a, p. 140).

The passage just cited suggests that we cannot plausibly postulate a sudden leap from a purely mechanical agency (a quasi-osmotic copying of behavior), to the fully-fledged cultural significations of the sacred.

Girard does not want a model based on a mysterious 'rupture' or leap. But nor does he want a gradualist model where only continuity is stressed. Yet the passage just cited seems to move in the latter direction. When does the hominid ever arrive at the sacred? When do concepts, representations, and symbolic misrepresentations finally follow upon an act of collective murder? It is not clear how reference to 'a very long time' matters, given that the murders which finally do generate the representational meanings and concepts of the sacred must also be the result of the non-representational mimesis if the latter is to be truly 'originary' — that is, if the anthropology is to remain 'rooted in the animal'. The only rupture for Girard is the collective murder, which he finds missing in those animal rituals which display the same mechanisms which, when carried one step further, produce the originary events (Girard, 1978a, p. 107). Thus, the same creature that is driven to participate in a collective murder without representational motivation must be the creature who will experience the fully-fledged significations of the earliest sacred representations. But how is this possible?

A potential solution to the problem is to point to a concurrent development of representational capacities alongside the increase in more mechanical forms of mimesis. On its own, the theory of an externally prompted and rather mechanical mimesis offers an incomplete explanation of complex beliefs and delayed, symbolic forms of modeling because different processes are at work in them (Bandura, 1986, p. 76). In other words, the increase in hominid brain size may very well have resulted in an increase in appropriative mimesis, but not *only* in it if this increase is to have produced organisms capable of sacred beliefs as well as actions motivated by these beliefs. The error would be to think that Girard's stress on the increase in mimesis must be taken as an exhaustive characterization of the impact of the increase in brain size on the hominids in question — as if the system of motivation determined by mimesis were the only behavioral system in the organism. At the same time as the tendency toward mimesis in certain types of situations increases, what emerges is an increased capacity for simulation or modeling of events in the environment. What we need to imagine is a creature who is in many ways already very intelligent and inventive. This is a creature whose interaction with the physical environment is not comprised of a highly limited repertory of motor responses generated by a purely mechanical form of agency (for the definition of mechanical agency, see Bandura, 1986, pp. 12ff.). As Bandura puts it, "A theory that denies that thoughts can regulate actions does not lend itself readily to the explanation of complex human behavior" (Bandura, 1986, p. 15). The hominid in question, then, already has well-developed capacities for memory, abstraction, forethought, intentional and purposive action. It probably has a tool-making capacity as well as a complex ability to orient itself in its natural environment, to undertake prolonged activities requiring planning and memory. There is evidence, for example, that Homo erectus (lower pleistocene, 1 million years ago) transported raw materials for stone axes from sites where they were available to the base camps where they would be used, the distance ranging as far as 60 kilometers. Gowlett infers from these data that these creatures must have had 'mental maps' of fairly extensive geographical areas as well as a capacity for planning and foresight over at least a day or more (1984, pp. 174—75). Such capacities outstrip a stimulus-driven form of mechanical agency. Yet we may at the same time imagine this creature as being prone to what may appear to us as a form of motivational rigidity in certain of its interactions with its fellow creatures; the rigid and rather mechanistic patterns that Girard attributes to appropriative mimesis at times totally govern its interaction with its fellows. One way to put this would be to say that what comes last is a capacity for higher-level social representations, such as beliefs about the nature of the relations in which the organism is engaged, and of course, a reflexive concept of self and a representation of the group as a whole. After all, the most intelligent creatures, the inventors of extremely sophisticated representational systems, frequently rely, in a highly unreflective manner, upon very basic and repetitive patterns of social interaction (examples would be the rigid and automatic nature of certain fairly common forms of joking behavior, or the interactions prompted by such emotions as anger). Why should the earliest hominids not also combine a complex, instrumental knowledge of features of the environment with a social order that is in some ways highly rigid and involving little self-knowledge? Would we not in this regard often resemble them?

## 7. TYPES OF MIMESIS

We recommend, then, a modified approach to the development of cognitive capacities in the process of hominization, namely, one that does not view the higher cognitive faculties as being generated or determined by primitive and mechanical forms of interactional mimesis. Instead, we suggest that behavior emerges from the complex interplay of different cognitive, motivational, and affective subsystems. Given this stipulation, it is possible to return to the question of the specific nature of the kinds of primitive mimesis discussed by Girard. Roughly, Girard's theory of primitive mimesis can be developed in the following three steps (we cannot undertake to provide more than a sketch of an argument in the present context; for more details, see *Models of Desire* (forthcoming).

(1) A detailed theory of non-representational mimesis should be provided. Roughly, this mimesis is a matter of what psychologists, following Allport, call social facilitation (Allport, 1920, 1924; Aronfreed, 1969; Zajonc, 1965; cf. the theory of motor mimicry proposed in Bavelas *et al.*, 1988). Under this description, basic mimesis involves no learning and requires that the organism already have a disposition to engage in the type of behavior that is matched.

(2) The results of (1) must be conjoined with a theory of what Girard refers to as 'appropriative' behavior, the goal being an analysis of the disposition for appropriative mimesis. Briefly, the class of appropriative behaviors includes only those behaviors that have an implicit goal of acquiring or maintaining exclusive possession or use of some object, place, or situation. Insofar as objective possibilities are concerned, the objects sought after in such behavior might very well be subject to division or to some form of cooperative use, but the organism's relation to them is such that these possibilities are excluded from the implicit aims of appropriative behavior. If we refer to the appropriative type of behavior as $\boldsymbol{\alpha}$ and to instances of it as $\alpha$, $\alpha'$, $\alpha''$, . . . , we may propose the following schematic formulation:

BASIC APPROPRIATIVE MIMESIS

Given organisms, $O_1, \ldots, O_n$, together in a single spatio-temporal location; a type of appropriative behavior engaged in by them, $\boldsymbol{\alpha}$, manifested in particular behaviors $\alpha$, $\alpha'$, $\alpha''$, . . . , a basic episode of appropriative mimesis requires that:

(a) At $t$ $O_1$ perceives $O_{1+n}$ do $\alpha$.

(b) At $t$ $O_1$ has a disposition to $\alpha$, this disposition being supported by a basic motivational system in $O_1$.

(c) At $t+1$ (just after $t$) $O_1$ does $\alpha'$.

(*Remark*: It must be taken for granted here that $\alpha$ and $\alpha'$ done by both organisms are directed toward the acquisition of one and the same object.)

(3) Moving from an individual to an interactional level of analysis, it must be shown how the interaction of more than one agent having a disposition for appropriative mimesis can produce the kinds of patterns called for in Girard's scenario, namely, the violent crisis and its violent resolution. More specifically, it must be shown how appropriative mimesis plausibly explains the emergence of aggression between intraspecifics and its escalation into a generalized crisis of violence.

For example, it could be argued that the progression to rivalry requires that both organisms not only be disposed to imitate each other's attempts to appropriate an object, but also that they persist in their effort to appropriate it. Scrambling behavior could be followed by hot pursuit, which could in turn lead to a sustained confrontation between two parties 'intent' on having the object. At this point behavior directly related to the retaining or seizing of the object might be replaced by gestures having the function of discouraging or incapacitating the opponent. Here one may introduce the rather plausible assumption that certain forms of violent behavior, appropriate to both attack and defense, are already part of the animal's behavioral repertory; that they should be set in motion in the kind of situation just described is hardly mysterious. It is not enough, of course, to say that this kind of sequence is realized when the 'instinctual bonds' are loosened, which implies that some powerful violent urge was already lurking in the breast of the beast. A more positive characterization is required; for example, it could be argued that the key element in the progression just sketched is the organisms' *persistence* in wanting and seeking to acquire the disputed object. The intensity of the motivational state that incites the organisms to want the object in the first place, as well as the appropriative nature of the want, lead to a situation where the organisms are drawn into conflict. Violent behavior ensues.

Even should the foregoing sketch be filled in convincingly, further work is required to explain the transition to a general crisis of violence involving the whole group. It is unclear that appropriative mimesis alone is up to the task, for Girard's 'violent crisis' is surely not a matter of a bunch of apes all trying to acquire the same banana. In a passage that is often overlooked by his commentators, Girard seems to respond to this issue by positing the existence of another type of mimesis, namely, *la mimesis de l'antagoniste* or antagonistic mimesis. This is a type of imitation that serves to explain the transition from a situation of hominids fighting for the possession of an object to one where they are engaged in a *generalized* imitation of aggression, that is, one where the imitator copies neither the specific goals nor the particular targets of the model's aggressive behavior. Thus he writes:

> Where there is no longer any object, there is no longer any appropriative mimesis in the sense we have given to this term. The only domain upon which mimesis can operate is the antagonists themselves. What happens, then, in the crisis, are mimetic substitutions of antagonists.
>
> If *appropriative mimesis* creates division by making two or more individuals converge on one and the same object, *antagonistic mimesis* necessarily brings them together by making two or more individuals converge on a single adversary that they all want to beat (1978a, p. 35).

Thus Girard posits another subcategory of mimesis that takes its place alongside the appropriative variety. Yet the implications for his attempt to ground hominid rivalry and crises of violence *in the object* need to be addressed. It may be objected, for example, that the postulation of antagonistic mimesis was made necessary by the impossibility of generating a generalized crisis of violence and a collective murder out of appropriative mimesis alone. In other words, the link between imitation and aggression had to be transformed to make the theory work: instead of imitating the other agent's attempt to acquire some object, the creature propelled by antagonistic mimesis imitates a more general behavioral schema, an aggressive pattern in which the target is really quite secondary. This gives the imitation of aggression a plasticity and volatility in keeping with the needs of the theory, but it also removes the mimetic process from its former roots in primary incentive objects. What is more challenging, then, is the problem of understanding the complex combinations of these different dispositions and capacities of the earliest members of culture.

The next step in the argument is the building of *detailed models* of the patterns of violent interaction in the hominid group, as well as of the group's convergence on the form of violent unanimity that Girard associates with scapegoating. A crucial step in the creation of such a model is the analysis of the various manners in which aggressive behavior may be imitated, as well as the different types of interactional patterns that result from different initial conditions and different manners of imitating aggression. Along these lines, the best guideline is provided by the stochastic models proposed by Orléan in his analysis of mimesis and market phenomena (1985, 1986a, 1986b, and 1989), although it is clear that important disanalogies emerge when one tries to apply these models to the problem of creating realistic scenarios of violent interaction in a hominid group.

*CREA, Ecole Polytechnique, Paris*
*McGill University*

## REFERENCES

Allport, Floyd H.: 1920, 'The Influence of the Group upon Association and Thought', *Journal of Experimental Psychology* **3**, 159—182.

Allport, Floyd H.: 1924, *Social Psychology*. Boston: Hougton Mifflin.

Aronfreed, Justin: 1969, 'The Problem of Imitation', In *Advances in Child Development and Behavior*. Vol. 4. Lewis P. Lipsitt & Hayne W. Reese (Eds.). New York: Academic, pp. 209—319.

Bandura, Albert: 1986, *Social Foundations of Thought and Action: A Social Cognitive Theory*. Englewood Cliffs, N.J.: Prentice-Hall.

Bavelas, Janet Beavin, Black, Alex, Chovil, Nicole, Lemery, Charles R., and Mullett, Jennifer: 1988, 'Form and Function in Motor Mimicry: Topographic Evidence that the Primary Function is Communicative', *Human Communication Research*, **14** (3), 275—300.

Boyd, Richard N.: 1973, 'Realism, Underdetermination and a Causal Theory of Evidence', *Noûs*, **7**, 1—12.

Boyd, Richard N.: 1979, 'Metaphor and Theory Change', In *Metaphor and Thought*. Andrew Ortony (Ed.). Cambridge: Cambridge University Press, pp. 356—408.

Boyd, Richard N.: 1983, 'On the Current Status of the Issue of Scientific Realism', *Erkenntnis*, **19**, 45—90.

Boyd, Richard N.: 1985a, 'Lex Orandi est Lex Credendi', In *Images of Science*. Paul Churchland (Ed.). Chicago: University of Chicago Press, 3—34.

Boyd, Richard N.: 1985b, 'The Logician's Dilemma: Deductive Logic, Inductive Inference, and Logical Empiricism', *Erkenntnis*, **22**, 197—252.

Fox, Robin (Ed.): 1975, *Biosocial Anthropology*. New York: Wiley.

Girard, René: 1972, *La Violence et le sacré*. Paris: Grasset. Trans. Patrick Gregory. *Violence and the Sacred*. Baltimore: The Johns Hopkins University Press, 1977.

Girard, René: 1978a, *Des Choses cachées depuis la fondation du monde*. Paris: Grasset. Trans. Stephen Bann & Michael Metteer. *Things Hidden Since the Foundation of the World*. Stanford, Ca.: Stanford University Press, 1987.

Girard, René: 1978b, 'Interview', *Diacritics*, **8**, 31—54.

Girard, René: 1978c, 'Delirium as System', *To Double Business Bound': Essays on*

*Literary, Mimesis, and Anthropology*. Baltimore: The Johns Hopkins University Press, pp. 84—120.

Girard, René: 1982, *Le Bouc émissaire*. Paris: Grasset. Trans. *The Scapegoat*. Baltimore: The Johns Hopkins University Press, 1986.

Girard, René: 1985, *La Route antique des hommes pervers*. Paris: Grasset. Trans. Yvonne Freccero. *Job: The Victim of his People*. Stanford, Ca.: Stanford University Press, 1987.

Gowlett, John A. J.: 1984, 'Mental Abilities of Early Man: A Look at Some Hard Evidence', In *Hominid Evolution and Community Ecology: Prehistoric Human Adaptation in Biological Perspective*. Robert Foley (Ed.). London: Academic Press, pp. 167—192.

Hinde, Robert A.: 1983, *Primate Social Relationships: An Integrated Approach*. Sunderland, Ma.: Sinauer.

Lacoue-Labarthe, Philippe: 1978, 'Mimesis and Truth', *Diacritics*, **8**, 10—74.

Livingston, Paisley: 1988a, *Literary Knowledge: Humanistic Inquiry and the Philosophy of Science*. Ithaca and London: Cornell University Press.

Livingston, Paisley: 1988b, 'Demystification and History in Girard and Durkheim', In *Violence and Truth*. Paul Dumouchel. (Ed.) Stanford: Stanford University Press, pp. 113—133.

Livingston, Paisley: (forthcoming), *Models of Desire: René Girard and the Psychology of Mimesis*.

Lockard, John S. (Ed.): 1980, *The Evolution of Human Social Behavior*. New York: Elsevier.

McKenna, Andrew J.: 1987, 'Supplement To Apocalypse: Girard and Derrida', this volume.

Madsen, Karl B.: 1974, *Modern Theories of Motivation: A Comparative Metascientific Study*. New York: Wiley.

Maxwell, Mary: 1984, *Human Evolution: A Philosophical Anthropology*. London: Croom Helm.

Montagu, M. F. Ashley (Ed.): 1962, *Culture and the Evolution of Man*. New York: Oxford.

Orléan, André: 1985, 'Monnaie et spéculation mimétique', In *Violence et vérité*. Paul Dumouchel (Ed.). Paris: Grasset, pp. 147—158. Trans. Mark Anspach: 'Money and Mimetic Speculation', In *Violence and Truth: On the Work of René Girard*. Paul Dumouchel (Ed.). Stanford: Stanford University Press, 1988, pp. 101—112.

Orléan, André: 1986a, 'La Théorie mimétique face aux phénomènes économiques', In *To Honor René Girard*, Alphonse Juilland (Ed.). *Stanford French Review*, **10**, 121—134.

Orléan, André: 1986b, 'Mimétisme et anticipations rationelles: une perspective keynésienne', *Recherches Economiques de Louvain*, **52**, 45—66.

Orléan, André: 1989, 'Mimetic Contagion and Speculative Bubbles', *Theory and Decision*, **27**, 63—93.

Potts, Richard: 1984, 'Hominid Hunters? Problems of Identifying the Earliest Hunter/Gatherers', In *Hominid Evolution and Community Ecology: Prehistoric Human Adaptation in Biological Perspective*. Robert Foley (Ed.). London: Academic Press, pp. 129—66.

Satinoff, Evelyn and Teitelbaum, Philip (Eds.): 1983, *Motivation*. New York: Plenum.

Snowdon, Charles T.: 1983, 'Ethology, Comparative Psychology, and Animal Behavior', In *Annual Review of Psychology.* Vol. 34. Mark R. Rosenzwweig and Lyman W. Porter (Eds.). Palo Alto, Ca.: Annual Reviews, pp. 63—94.

Sober, Elliott (Ed.): 1983, *Conceptual Issues in Evolutionary Biology.* Cambridge, Mass.: M.I.T. Press.

Toates, Frederick: 1986, *Motivational Systems.* Cambridge: Cambridge University Press.

Toates, Frederick M. and Halliday, Timothy R.: 1980, *Analysis of Motivational Processes.* New York: Academic.

Washburn, Sherwood L.: 1962, 'Tools and Human Evolution', In *Culture and the Evolution of Man.* Ashley Montagu (Ed.). New York: Oxford, pp. 13—19.

Wilson, Peter J.: 1980, *Man, the Promising Primate.* New Haven: Yale University Press.

Zajonc, R. B.: 1965, 'Social Facilitation', *Science,* **149**, 269—274.

# PART II

# THE ORIGIN OF MONEY: SYMBOLS AND TEXTS

ANDRÉ ORLÉAN

# THE ORIGIN OF MONEY

Translated by Paisley Livingston, McGill University, Montréal, Québec

## INTRODUCTION: COLLECTIVE REALITIES AND INDIVIDUAL STRATEGIES

Most economists now agree that the basis of money is its general acceptance, as a medium of exchange, among the members of a given community. Although the metallist tradition long dominated economics, modern thinkers no longer hold that the acceptance of money as a medium of exchange should be understood as resting on a belief in the intrinsic value possessed by money itself. When Frank Hahn sought to lay the groundwork for an adequate theory of money, his first statement was that he was "interested in an economy where money is of no intrinsic worth and is universally accepted in exchange" (Hahn, p. 1). From this perspective, any reference to gold as a standard of value is taken as a failure to have reached a true understanding of monetary phenomena. In his famous textbook, Paul A. Samuelson goes so far as to make the following declaration: "From the standpoint of understanding the nature of money, it is perhaps simpler that the citizen's gold certificates and coins no longer exist. The modern student need not be misled as were earlier generations of students, by some mystical belief that 'gold backing' is what gives money its value. Certainly gold, as such, has little to do with the problem" (Samuelson, 1976, pp. 278–279). Modern monetary theory, then, is concerned solely with 'fiat money', and as Neil Wallace comments, "there are two widely accepted characteristics of fiat money: inconvertibility and intrinsic uselessness" (Wallace, p. 49).

Although this modern perspective may seem to be quite clear and simple, it nonetheless encounters a major difficulty: if money has no intrinsic usefulness why do agents accept it? And if they only accept it on the condition that others do, what makes them confident that this condition will be widely satisfied? In other words, what is the nature of the obligation that forces other parties to exchange their goods and services for a sign having no intrinsic value? This is the central question for those who are interested in the origin of modern or 'fiat' money.



Francisco J. Varela and Jean-Pierre Dupuy (eds), Understanding Origins 113–143.

ANDRÉ ORLÉAN

# THE ORIGIN OF MONEY

(Translated by Paisley Livingston, McGill University, Montreal)

## INTRODUCTION: COLLECTIVE REALITIES AND INDIVIDUALIST STRATEGIES

Most economists now agree that the basis of money is its general acceptance as a medium of exchange among the members of a given community. Although the metallist tradition long dominated economics, modern thinkers no longer hold that the acceptance of money as a medium of exchange should be understood as resting on a belief in the intrinsic value possessed by money itself. When Frank Hahn sought to lay the groundwork for an adequate theory of money, his first statement was that he was "interested in an economy where money is of no intrinsic worth and is universally accepted in exchange" (Hahn, p. 158). From this perspective, any reference to gold as a standard of value is taken as a failure to have reached a true understanding of monetary phenomena. In his famous textbook, Paul A. Samuelson goes so far as to make the following declaration: "From the standpoint of understanding the nature of money, it is perhaps simpler that the citizenry's gold certificates and coins no longer exist. The modern student need not be misled as were earlier generations of students, by some mystical belief that 'gold backing' is what gives money its value. Certainly gold, as such, has little to do with the problem" (Samuelson, 1976, pp. 278–279). Modern monetary theory, then, is concerned solely with 'fiat money', and as Neil Wallace comments, "there are two widely accepted characteristics of fiat money: inconvertibility and intrinsic uselessness" (Wallace, p. 49).

Although this modern perspective may seem to be quite clear and simple, it nonetheless encounters a major difficulty: if money has no intrinsic usefulness, why do agents accept it? And if they only accept it on the condition that others do, what makes them confident that this condition will be widely satisfied? In other words, what is the nature of the obligation that forces other parties to exchange their goods and services for a sign having no intrinsic value? This is the central question for those who are interested in the origin of modern or 'fiat' money.



*Francisco J. Varela and Jean-Pierre Dupuy (eds), Understanding Origins,* 113–143.

To the extent that it is identified with a right that holds equally well for all commodities possessed by all parties in exchange, money presupposes that there is a *generalized* agreement among all members of a given community. This point is stressed by the many authors who analyze the acceptance of money as arising from certain types of behavior that express 'the universal agreement of mankind'. Raymond Barre insisted on the notion of *confidence*, which he holds to be "inseparable from the very idea of a community of payment." François Simiand stresses the fact that "all forms of money imply a social belief and *faith*." Friedrich Von Wieser emphasizes the idea that '*habit*' is the real basis of the 'massive acceptance' of money (cited in Barre, pp. 323—325). Once these behavioral dispositions are in place, the acceptance of money is no longer a problem, for these dispositions can be identified as what causes all agents to accept money. Since all agents act the same way, the beliefs that are the basis of their behavior are validated, and the generalized agreement that was presupposed is effectively realized. But what justifies the behavior?

Common sense suggests that it is a matter of a suspension of our faculty for engaging in individual critical thinking: we follow the monetary rule out of habit, confidence, or faith in a mechanical form of subservience that is by no means the product of some form of rational calculation. The holist approach interprets this temporary lapse of reason as the effect of a social force that imposes its own ends on the will of the individual, and this social force is *society itself as an autonomous entity*. According to the holist perspective, the community is not reducible to a simple statistical aggregation of individual behaviors, but has a specific identity that stands above the individual and gives priority to collective ends, including some ends that thwart individual desires. This sort of holist approach has mainly been used to account for the manner in which traditional societies understand social order. Within traditional, holistic societies, the social order was thought to be founded on a series of past events and agents having a sacred or religious status; an immemorial tradition is held to be beyond the grasp of all individuals, who can only obey and embody its structures. Such a society is instituted by means of a basic rift between the time of the origin and the time of present, strictly human actions. As a consequence of this type of institution, the rules and modes of functioning of the social order are placed out of the range of human criticism. This means that society has a radical alterity in relation to its members, for part of the social being is withdrawn from the members of the group and has a specific identity of its own. This kind of holist view of social relations claims to provide a satisfactory image of the monetary phenomenon, an image in which money is a manifestation of an essentially collective reality, namely, the unanimity of the members of society, a unanimity which, in imposing certain beliefs on the agents, alters their behavior and thereby guarantees its own stability. This is the conception that is expressed by anthropologists when they write that "There is no money in the absence of a transcendent order that gives it the quality of being a materialization of the social totality" (Barraud *et al.*, p. 507).

This holist conception of money evokes an image of modern market societies that contrasts rather sharply with many of our most firm beliefs, particularly the idea that "the order that presides over the communal life of mankind is no longer a given, something that people receive from a source outside of the purely human sphere" (Gauchet, p. 249). Contemporary thinkers disagree with holism, then, and claim that social order has no radical alterity because it is created by men and women. Individualism is the name for the set of values that "valorizes the individual and neglects or subordinates the social totality" (Dumont, p. 264), holding that "there is no longer anything ontologically real above the particular being" (Dumont, p. 73). With the advent of individualist ideology there emerges a conception of social institutions that denies the legitimacy of those archaic forms of human relations that amount to blind confidence and habitual obedience. Such a conception of human relations finds a precise and highly elaborate formal expression in economics, or at least in a large part of the work produced by that discipline. The economic tradition considers the basic social relation to be the contract, freely negotiated between rational subjects with their own interests in view. Insofar as it is based on the agents' conscious agreement, the contract is the paradigmatic expression of individualist values. But the question that must be dealt with in the present context is whether the individualist approach in fact constitutes a viable alternative to holist understandings of money: can the fact of money really be explained as the result of rational calculations and agreements? To put the issue in a more polemical way, we may ask whether money is in fact wholly in step with individualist values. Such questions are made all the more difficult to answer by the fact that it is a matter of trying to conceive of the presence of collective realities from within a world view that is allergic to any and all holist categories.

Political economy as a whole bears witness to the problem, which finds a first expression in the fact that money, as the entire society's writ, takes a strange form: "money appears as a bill of exchange from which the name of the drawee is lacking" (Simmel, p. 177). If money happens to fall outside the central economic category, the private contract, this is because in individualistic societies the community no longer has any specific identity of its own, and it is *impossible to form a contract with it.* It is only a metaphor to speak of money as 'a bill of exchange' because there is no one who bears the obligation of repayment. Put in this manner, the question of money is identical to a problem that individualist strategies always encounter, which is that of accounting in purely individual terms for collective entities, and more specifically, for rules that have a coercive effect on individuals. How can such collective realities be reconstructed entirely in terms of the logic of interactions between private subjects? Can all institutions be reduced to contractual arrangements? The problems that such strategies encounter are well known, but the stakes remain fundamental, for it is a matter of demonstrating that individualist values remain viable and that it is possible to conceive of a stable and coherent social order rid of holist vestiges. Can the individual's adherence to a collective order be explained as the result of a rational calculation? Is money reducible to contractual relations? The specificity of an economic approach to modern money has its basis in a fundamental decision that rules out searching for an origin of money in some kind of transcendent referent, such as gold, law, habit, faith, or confidence, standing outside private exchange. Instead, the origin must be sought in a form of self-organization. *The reductionist individualist strategy can be defined as the attempt to analyze social exteriority as self-institution.* But can this be done? Has economic theory managed to explain the reality of money entirely in terms of individual calculations? The first part of this essay will be devoted to the latter question, and a negative answer will be defended: *money reveals the inadequacy of the individualist order.* This thesis has many implications for our understanding of the relation between archaic and modern forms of money. It implies a reversal of perspectives, a reversal that corresponds to the analyses of certain anthropologists: "we need not ask whether archaic money matches various possible definitions of modern money, but whether archaic money does not reveal some unexpected features of the modern form. It seems to us that archaic and modern money are both, first and foremost, money" (Barraud *et al.*, p. 507). The second part of this essay will analyze the relations between archaic and modern forms of money, which will give us a better understanding of the origin of the modern use of money.

## 1. MONETARY INCOMPLETENESS, OR THE IMPOSSIBILITY OF TRANSPARENCY

The most widespread conception of money in economics is the quantity theory of money. An examination of this theory fully reveals the kinds of problems that the discipline encounters when it must deal with monetary realities. We will see that individualist values and money are only reconciled in this theory by means of quasi-disappearance of the fact of money.

### 1.1. *The Disappearance of Money*

Money has a very special status in quantity theory, as can be shown by adopting a dichotomy suggested by Samuelson, who proposes that monetary phenomena are said to have qualitative as well as quantitative dimensions. In terms of the former, money is essential, for it is the founding institution of market economies — economists having demonstrated at great length that barter systems are inefficient. Samuelson comments that "once this qualitative advantage had been realized by the adoption of market structures using *M*, the *quantitative* level of *M* was of no particular significance" (1968, p. 172). In keeping with this opinion, economists since John Stuart Mill have traditionally thought of money as the "lubricant of industry and commerce". In order to identify this sort of indifference to the quantity of money, we may speak of the *neutrality* of money. The basic idea here is that neither the structure nor the level of production and exchange is altered by variations in the quantity of money issued, for both depend entirely on the real economic factors, namely, tastes, endowments and production possibilities. Variations in the supply of money only modify the general level of prices. Following Don Patinkin's analysis, we may characterize the neutrality of money as the fact that neither relative prices, exchanges, nor production is affected by the transformation of monetary economy into a barter system: as far as the determination of real economic variables is concerned, the economy looks as if it is moneyless. What comes to mind here is an analogy to language, for to the extent that language makes it possible to express our beliefs without disturbing their meaning, we have the same duality: on the one hand the important 'qualitative' dimension (making communication possible), and on the other hand, the neutrality of a medium in relation to the 'real' variables (the meanings transmitted in the language). In other words, if money is neutral, this is because it is only a pure form having no effect on the constitution of relations between individuals, which depend entirely on the economy's real variables. According to such a conception, the existence of money cannot be put in question because it never restricts the freedom of the agents on the market. It is never a source of conflict and is only a simple tool that can be used by everyone.

Such an analysis may suggest that the monetary phenomenon involves a certain dualism and ambivalence, but there is no paradox here. In fact, if money is both indispensable and neutral, these two characteristics have to do with two separate moments in time. The first corresponds to the constitution of the market sphere. Here money is absolutely essential because it makes possible the creation of a whole new universe. But in its synchronic dimension, money wipes away every trace of this origin and becomes a mere tool. This temporal dichotomy makes it possible to avoid the paradoxes of self-institution: the social and collective quality of the monetary sign is completely caught up in the genetic moment. Once it has been produced, this quality appears to the agents as a simple datum, a resource that stands ready to be used. Money is worth accepting, then, because of the services that it renders.

To round out this presentation of the quantity theory, we must reflect on the way in which the services that money renders have been formulated. Typically, three motives for holding money are identified: the precautionary-motive, the speculative-motive, and the transactions-motive. Conceptualy it is the latter that has priority because it embodies the primary characteristic of monetary economies, the multilateral nature of exchanges. Patinkin presents the desire to engage in transactions as being linked to the absence of a synchronization of payments made and received, which in turn gives rise to a desire for liquidity, which would diminish the inconvenience caused by temporary insolvency. According to this conception, money has a *direct utility*. This makes money different from titles, that, in general, only have an indirect utility: titles only have the pro rata value of what they make it possible to buy. It follows that money appears in agents' utility functions on the same footing as commodities. The specific service that it renders is liquidity. Money is a kind of insurance that agents take out against the risk brought by uncertainty concerning the payments they will be receiving in the future. In this manner, the demand for money has the same properties of any other demand for a commodity. Its stability is that of the need that it satisfies. In other words, once money is present, we can consider it to be like any ordinary commodity, for it is simply the commodity that satisfies the desire for liquidity. But such a conception makes the social nature of the service that money renders disappear. Money's objective status is taken as having been established once and for all; the desire to engage in transactions can be satisfied by using whatever money one happens to have in one's possession.

Such a theory does not concern itself with the kinds of questions with which we began this essay. It is simply laid down as a matter of fact that everyone accepts money because it has a utility for each and every individual. That this utility depends finally on everyone else's acceptance of the monetary sign is not deemed pertinent, for it is held to be a purely historical matter. What the theory maintains is that once money has been created, it will always be accepted by agents because it satisfies an objective need. Such an understanding of money makes possible a very particular solution of the problems raised by money's status as a collective object. Quite simply, these problems are made to *disappear*. The theoretical disappearing act that is in question here is double. First, it consists in making the formation of money a purely historical matter. But this move can hardly succeed when the very process of exchange brings back to the fore what has been relegated to history, namely, the question of the 'acceptability' of money. Thus a second disappearing act must follow the first so that no doubts about monetary 'quality' can arise as a result of exchange. Money's second disappearance is brought about by means of the concept of neutrality. Neutrality means that money is involved neither in the determination of interpersonal relations, nor in the constitution of individual preferences. In other words, money is a means that is wholly external to the circulation of commodities. The double disappearance of money makes it impossible to raise any specific questions about the social nature of money. In such a theoretical framework, the presence of money is in no way incompatible with the rational calculations of individuals, their maximizing of personal utility. Nor does the acceptance of money require any sort of blindness, irrational faith, or habitual compulsion. Its existence does not contradict an individualistic understanding of social relations.

Although its elegance cannot be doubted, this theory hardly seems to be entirely satisfactory. The problems that money raises cannot be solved by simply making them disappear. Nor can we assume that monetary 'qualities' are required once and for all, or that the process of exchange, itself motivated by agents' desires to engage in transactions, will reproduce these qualities in a wholly unproblematic way. Monetary crises in their many different forms remind us that money is not immortal. The market society constantly engages in a more or less latent questioning of its money, measuring its ability to serve as an impartial means of exchange. What form does this questioning take? What are its determinants? Following a tradition that embraces such prestigious theoreticians as Karl Marx and John Maynard Keynes, we believe that the key to these questions is the use of money as a means of hoarding.

### 1.2. *Hoarding and Critical Reason*

Marx suggests that we only get a partial and highly fragmentary picture of money when we focus on its use as a measure of value and as a medium of exchange, two functions to which he referred respectively as ideal and symbolic money. "Gold is capable of being replaced by tokens that have no value in so far as it functions exclusively as coin, or as the circulating medium, and as nothing else. . . . Its functional existence absorbs its material existence" (Marx, 1936, pp. 144—145). But in the phenomenon of hoarding (*die Schatzbildung*), real money appears. It reveals itself as "the sole form of value, the only adequate form of existence of exchange-value, in opposition to use-value, represented by all commodities" (Marx, 1936, p. 146). Marx knows to what extent hoarding can endanger the harmony of market relations, for it threatens the very continuity of exchange. By dissociating acts of sale and purchase, hoarded money no longer serves as a simple tool that is at everyone's disposition. Instead, it becomes the locus of a power that every individual can wield in order to affect the circulation of goods. The perfect order imagined by the quantitative theory of money is by the same stroke profoundly disturbed. Here we see the return of what the concept of neutrality repressed, namely, money as a social force: in the case of hoarding, "social power becomes a private power of private individuals" (Marx, 1883, p. 105; 1936, p. 149). The danger caused by this sort of behavior is all the more great because "in its very nature the desire (*Trieb*) to hoard is measureless" (Marx, 1883, p. 106; 1936, p. 149). Unlike the demand for money, the desire to hoard is highly unstable. Here individual arbitrariness finds its absolute form. For Marx, its logic is essentially a matter of psychology, and more specifically, a psychology of greed.

These reflections can be interpreted as follows: in circulation, individual rationality encounters no obstacle to the reproduction of the established conditions of the market, and in fact, individual rationality is the necessary basis of those conditions. Hoarding vitiates this situation. It expresses the possibility of idiosyncratic individual demands that violate the systematic nature of economic exchange. Another way to describe this shift is to note that hoarding causes money to cease being a neutral means: on the contrary, it is desired as such. The emergence of 'real money' destroys the harmony of market relations. But how can such a dynamic be understood? It seems that Marx does not fully develop his analysis. More specifically, the nature of the power that money provides, and the reasons why agents are consumed by the desire to hoard, are not made clear. The author who actually explains this behavior and its economic significance is Keynes.

In the Keynesian theory, the desire to hoard money is closely linked to a very particular conception of the temporality of the market: "our desire to hold money as a store of wealth is a barometer of our distrust of our own calculations and conventions concerning the future. . . . The possession of money lulls our disquietude" (Keynes, 1937, pp. 218–219). Our previous remarks about transactions also emphasized the crucial role played by uncertainty in the formation of the demand for the means to engage in them. But the kind of uncertainty about the future that Keynes has in view is very different from the sort discussed by Patinkin, for it is not simply a matter of an inability to foresee when payments will be received. In the world that Keynes envisions, the future is only a set of more or less plausible possibilities, the realization of which depends on human action. If at the present time a certain number of conventions give meaning to individual decisions, in the future these representations may change. As suggested by George Shackle, uncertainty is epistemic and is a function of the evolution of our knowledge. It follows that it is rational for agents to doubt the pertinence of the very beliefs that currently govern economic activity, for these beliefs are, like all human knowledge, essentially fallible. In an earlier text, we have underscored the affinity between Keynes's notion of uncertainty and a Popperian understanding of scientific knowledge (Orléan, 1989a).

Systematic doubt about our future knowledge may be quite insidious. It finds a privileged expression in the phenomenon of hoarding, a withdrawal or retreat from the circulation of commodities on the market. Thanks to his or her holding of money, the agent does not have to make any decision that would lead to a concrete investment. He or she views the world as a set of possibilities between which no choice is necessary. The agent is on hold and stands outside the game in an attitude of self-reserve. Saving money bestows flexibility and makes it possible to adapt to all future situations. Shackle elaborates on the same theme when he writes that "A money-using economy is one which acknowledges the permanent insufficiency of the data for rational choice. For money is the means by which choice can be deferred until a later and better-informed time" (Shackle, p. 160). Let us underscore the rather special nature of this sort of attitude. In hoarding doubt about the future takes its strongest form, and appears as a criticism of the existing values and beliefs, a criticism that need not be accompanied by any alternative hypotheses. This attitude gives perfect expression to the individualist values according to which the rational individual is someone who puts all received habits and beliefs in doubt. As Jacques Bouveresse notes, "the birth of rational man coincides with the realization of fallibility" (Bouveresse, p. 12). The practical instrument of this critical reason is money, as a store-of-value. In fact, private withholding of money is what allows individuals to express their doubts about the validity of predominant social conventions. Specifically individualist values correspond, then, to the emergence of money as a store of value, and to the interruption of the circularity of exchange that this entails. Here we see an essentially modern use of money, and in what follows we will return to this intuition in an analysis of archaic monetary phenomena. Yet let us note first that money as a store of value gives rise to a rather paradoxical situation. Does not hoarding depend on a rather high degree of confidence in the monetary sign itself, a confidence that is strangely wedded to a profound doubt concerning all of the concrete activities that are in fact the ultimate basis of money's worth? What would the value of money be if everyone abandoned all productive activity and began to hoard? Hoarding would seem to suggest that society's belief in the permanence of money can be partially autonomous. *Legitimacy* is the term we use to refer to the fact that money can be the object of a degree of confidence that surpasses the real conditions of the existence of the monetary order. Legitimacy is what authorizes the critical activity we evoked above. It attenuates the pressure to conform to both opinion and the constraints of production. It causes the emergence of a representation of time in which the future is a set of possibilities and innovations. But to what extent is legitimacy compatible with a situation where the agents are fully aware of the nature of the market? In such a context, what is the meaning of their confidence in money?

### 1.3. *Specularity and Legitimacy*

To answer our questions about the modern form of money, we must begin with an analysis of the individual behavior involved in hoarding. This problem will not be dealt with in detail here. Instead, we will reveal the logic of the difficulties that such an analysis encounters, showing how these problems affect our understanding of money. The basic argument runs as follows. The quantitative theory presents liquidity as an intrinsic property of money. Holding money is the product of a simple calculation whereby the individual determines what amount of liquidity he or she desires. Thus the demand for money only reflects personal inclinations. But in the case of the problems that we have in view, the situation is totally different. Hoarding money is not a simple effect of individual taste. Instead, *the individual demand for money depends on the behavior of other agents.* A process of this sort is not necessarily stable, as we shall see.

For any agent, *i*, that agent's acceptance of a worthless sign, money, in exchange for a commodity, depends on that agent's expectations about the future acceptance of the same sign by another agent, *j*. The particular qualities of the sign hardly matter, for what is essential in determining agent *i*'s decision is *i*'s expectation about the behavior of *j*. Agent *i* will only take the money if he knows that *j* will accept it in turn one day. But, to the extent that *j*'s acceptance of the money also depends on *j*'s expectations about a new agent, *k*, *i*'s acceptance of money depends on *i*'s expectations about *k*'s acceptance of the monetary sign. It should be obvious that this reasoning hardly stops with agent *k*. Thus, the acceptance of money depends on an infinite chain of expectations about the expectations of other agents. 'Specularity' is the name we give to this kind of situation where the agents' behavior is based on their reciprocal expectations about each other's behavior. The relevance of such an analysis consists in not considering as resolved what should, obviously, be the core of the monetary theory: what is the monetary quality founded on? This quality is no longer taken to be something that history has invented once and for all. Instead, the qualitative dimension of money is viewed as the product of specularity. Understanding the origin of money, then, is a matter of understanding how specularity is expressed; it is a matter of knowing how specular beliefs evolve and are stabilized.

Recent economic theory has taken these issues quite seriously (Orléan, 1989b). A wide range of models show that specular processes give rise to the emergence of unanimity: all individual opinions converge on a single belief. The content of the belief in question is indeterminate in the sense that a large number of opinions can become the object of unanimity. As soon as unanimity is achieved, the belief is self-validated, for, as specular logic would have it, the veracity of a belief depends essentially on its being shared by others. Let us emphasize the fact that this kind of self-validation is largely independent of the specific content of the belief. What we are dealing with here is a process similar to self-fulfilling prophecies: if everyone thinks that sign $X$, whatever it may be, is a store of value, then it becomes one since everyone accepts it; the initial belief is validated.

The convergence of individual beliefs on a single belief may be thought to be the 'normal' form of specularity. The community relates to itself in the form of an 'external' opinion shared by everyone, an opinion that can also be given the name of 'common sense'. This externalization leads to a particular form of specularity: to anticipate what the others are going to do, one need only refer to the common beliefs. Common knowledge of the single belief that has been made the object of unanimity takes the place of an endless chain of reciprocal guesses about what the others will think and do. Intersubjectivity no longer takes the form of wondering about the others. In such a configuration, everything occurs *as if* the monetary quality is a simple given that imposes itself on the system from the outside.

The approach that I have just evoked is quite interesting, for it denies that money has any direct utility. Consequently, money is not introduced into our description of the individual's utility functions, as was the case in Patinkin's approach. The fetishism that surrounds money in other approaches is totally demystified: far from being seen as the effect of certain intrinsic properties of a particular commodity, the monetary quality is viewed as the product of a specular intersubjective dynamic. It may be concluded, then, that this approach is the one that satisfies the requirements of the individualist ideology most fully. The theory of money that it advances is one in which money is something subjects produce themselves: money emerges as the result of individuals' efforts to transcend the limits of the barter system. Yet one question remains outstanding. The fact is that analyses of speculative dynamics, such as those of financial markets, show quite clearly that unanimous collective belief can fall apart if its only basis is individual calculation. In other words, the attitude that leads each individual to choose a store of value in function of an expectation about what others believe can indeed lead to unanimity about the value of a certain sign, but it remains to be shown whether this contingent unanimity enjoys any stability. The theory that I have just outlined reveals that the object of unanimity has a purely conventional nature: the means of saving is in fact indeterminate since many objects could play this role. According to a pure logic of rational calculation, the objects chosen are totally conventional. *The question that is raised, then, is that of knowing what effect knowledge of the conventional nature of a convention will have on its stability.* To put the question more precisely, we may ask whether the stability of the convention can be guaranteed by the agents' knowledge of the fact that they have to use conventional objects to solve their coordination problems. This is a central question. Two ways of responding to it can be identified, and to them correspond two different analytic perspectives. If we respond affirmatively (reasoning that the stability of the convention can be founded on the agents' knowledge of the conventional nature of the solution), then it is possible to believe that the legitimacy of money involves no misrepresentation. Agents accept the existing sign because they know that any other one could play the same role. It remains true that the historical emergence of this particular monetary sign involved a process of which the individuals are not fully aware, but this historical indeterminacy has no effect on their conduct. Everyone knows that history plays a role in the production of conventional objects. In such a context, the agents' cognition conforms to the real process that is the basis of the emergence of money: there is a complete social transparency.

The second analytical perspective, which is our own, contends that the knowledge of the conventional nature of the monetary sign is destructive insofar as it casts doubt on the object's ability to serve its function as a store of value. The same specular logic that under some conditions can generate unanimity can easily degenerate into a cumulative and self-validating process that destroys it. In other words, the stability of the convention requires a certain degree of misrepresentation taking the form of a 'materialization' of belief. Money requires a certain degree of opacity, a certain ignorance that expresses itself by means of references to 'external' objects. *The process of externalization by means of which social unanimity is rendered legitimate cannot be based on rational calculations alone.* There is a necessary gap between the beliefs and the formal model. Even if the agents are totally convinced that the rules have to be conventional, their theoretical awareness of that necessity is not enough to make any particular institution stable. This gap sets a limit to the project of transparency, and it is the very essence of legitimacy. Legitimacy necessarily mobilizes a lack of understanding of the very process that engenders it, and as a result, it inevitably involves a certain opacity. The opacity constitutive of legitimacy is what makes society exist as a totality that is partly disconnected from individual wills. Thus this opacity belongs to a mode of relations in which social forms are not reducible to individual calculations: here the critical perspective finds its limit. In saying this we are very close to Friedrich Hayek's analysis of institutions, for in his work, the notion of institution corresponds to those aspects of the monetary phenomenon that we have been emphasizing — the putting in question of the individualist project, the limitations of critical reason, and the opacity of social relations in regard to individual awareness. Traditionally, legitimacy has been expressed by means of forms representing an external guarantee — such as the gold standard. More fundamentally, it should be shown how the present approach makes it possible to understand liquidity as a fact that transcends individual calculations and that makes the adjustments of the market possible.

### 1.4. *The Impossible Dematerialization of Money*

To conclude this first section, let us recall Georg Simmel's interesting analysis of the same problem. Simmel vehemently opposed the metallist conception that was still widespread at the time, and in many lengthy passages in *The Philosophy of Money*, he defends the thesis that the functions of money in no way depend on a material basis or support.

On the contrary, he stresses the fact that only a dematerialization of money could allow it to realize its function fully. In the terms of the previous discussion, Simmel's 'dematerialization' of money is the same thing as the critical process entailed by individualism, a process that sheds light on the conventional nature of all the symbols that coordinate social life. However, after he has insisted that money is an ideal form, Simmel notes: "Yet money cannot cast off a residue of material value, not exactly for inherent reasons, but on account of certain shortcomings of economic technique" (Simmel, p. 158). And in another passage, he adds that the "elimination of the intrinsic value of money is impossible" (Simmel, p. 162). In other words, from a formal point of view, nothing prevents money from being a pure sign, but certain imperfections make it impossible for this process to reach its completion. What, then, are these imperfections? According to Simmel they are of two types, but only the second one is pertinent in the present context: "Although, in principle, the exchange function of money could be accomplished by mere token money, no human power could provide a sufficient guarantee against possible misuse. The functions of exchange and reckoning obviously depend upon a limitation of the quantity of money, upon its 'scarcity,' as the expression goes" (Simmel, p. 159). In other words, as a pure symbol that stands outside all empirical reality, money is so indeterminate and vague that it offers no resistance to arbitrariness. And it is the anticipation of this possibility more than the possibility itself that makes it impossible for money to become a purely abstract, totally disincarnated symbol, freed from all specific rules governing its issue. Thus it would appear that real money is never adequate in relation to its own concept, or to put it in our own terms, the market order never fully corresponds to the values of individualism. The gap in question, or the supplement that real money displays, is what we have defined as legitimacy.

These reflections lead Simmel to consider the evolution of money as a wholly paradoxical process, one having a goal that it can never reach because doing so would make it impossible: "It is not technically feasible to accomplish what is conceptually correct, namely to transform the monetary function into a pure token money, and to detach it completely from every substantial value that limits the quantity of money, even though the actual development of money suggests that this will be the final outcome" (Simmel, p. 165). Here there would seem to be a basic duality of the conceptual and the real, following which our intellectual representation of the process is by nature out of step with reality. According to Simmel "a great number of processes occur in the same manner; they closely approach a definite goal by which their course is unambiguously determined, yet they would lose precisely those qualities that led them towards the goal if they were actually to reach it" (Simmel, p. 165). Thus we may add that the monetary process necessarily involves a certain misrepresentation of the conditions of its own existence. A false representation of the process is necessary for the very existence of the process. Such is the nature of the opacity that the individualist monetary order requires in order to exist: the representation of the goals of the system must be false. The individualist order gives itself the goal of doing away with every sort of opacity, but it could never realize this goal without destroying itself. The illusion that is characteristic of individualism is its failure to understand that the goal it adopts is out of reach. But it does remain possible to consider this opacity as being the smallest amount that is compatible with the existence of society, at least in the sense that it takes the paradoxical form of a belief in the possibility of total transparency. Thus money reveals the incompleteness of individualist values.

## 2. MONEY IN HOLIST SOCIETIES

We are not in a position to provide a comprehensive synthesis of the vast literature on archaic forms of money — a literature characterized by a diversity of interpretations and by a multiplicity of particular examples. The use of ethnographic documents that will be made here has a rather different end, for it is a matter of seeing whether the distinction between money as a means of circulation and money as a store of value also applies to traditional societies. To what extent does this distinction correspond to two different conceptions of time, the one based on circularity, the other on innovation and the unforeseeable nature of the future? How do archaic societies understand legitimacy? Such will be the topics to be taken up in what follows.

Let us begin with a criticism of the economic fable according to which money was born as a result of the difficulties of a system based entirely on barter. This fable would have it that the movement from barter to money was part of a general evolution towards reason, an evolution in which the archaic nature of the first forms of exchange was replaced by more rational ones. To our knowledge, this fable has been rejected unanimously by everyone who has taken the trouble to study archaic societies. Karl Polanyi, for example, has vigorously demonstrated the stupidity of the economist's fable. According to him in premodern societies the economy is embedded within a very dense tissue of social obligations that control its operation and hold it within narrow bounds. This social structure may be very stable. Its own internal logic does not lead to any sort of irresistible extension or development of market relations: "The limiting factors arise from all points of the sociological compass: custom and law, religion and magic, equally contribute to the result, which is to restrict acts of exchange in respect of persons and objects, time and occasion" (Polanyi, p. 61). And Arthur M. Hocart writes that when economists and anthropologists pretend that the monetary system was created as an answer to the limitations of barter, they impute to primitive men the preoccupations of the modern financier. He claims that we must try to rid ourselves of modern prejudices and flatly asserts that his evidence suggests that the origin of modern metallic money is religious (Hocart, p. 108). These criticisms of economism are confirmed, for example, by the fact that only a small part of the monetary circulation in archaic cultures has to do with economic exchanges. When economists hear these criticisms, they usually fail to take them into account in their own analyses, which for the most part remain based on the idea that social life is identical to the combination of individual rational strategies. Yet social life is not in fact of this nature, and to continue assuming that it is amounts to what Polanyi calls "The habit of looking at the last ten thousand years as well as at the array of early societies as a mere prelude to the true history of our civilization which started approximately with the publication of the *Wealth of Nations* in 1776" (Polanyi, p. 45).

We, on the other hand, would follow Hocart in insisting upon the importance of exploring the hypothesis that money has a religious origin. *Such is the source of its legitimacy.* More specifically, we would stress the role played by sacrifice within the earliest religions. Our comments on these matters are largely speculative, even if they are based on a large number of examples. What follows, then, is a matter of hypothesis. Our reflections should be understood as a series of questions posed to the specialists in archaic forms of money.

### 2.1. *Archaic Money and Circular Forms*

A first hypothesis concerns the role of circular forms of the transmission of money. These forms are a perfect expression of the holist idea of society as an organic totality. In a process that has neither beginning nor end, everyone is linked to everyone else by means of a complex network of reciprocal gifts of money. Money has a value only in relation to this totality that it actualizes and creates at the same time. According to William H. Desmonde, the story of King Arthur and the Knights of the Round Table reveals that the source of monetary value resides in "the feelings of mutual devotion and fellowship and in the bonds of loyalty among the communicants" (Desmonde, p. 19). The legend includes two details that would appear to be fundamental symbols of social unity: the circular form of the table, and the banquet, "a ritual communion meal in which the shared food symbolized mutual dedication among the communicants" (Desmonde, p. 21). The Holy Grail symbolizes the group's force, its cohesion, as well as a feeling of powerful emotional unity. Desmonde believes it to be the fundamental form of money.

Louis Gernet takes up the problem of the origin of money and of value in ancient Greece, studying a series of objects called *agalma*, which are associated with notion of wealth and value. The *agalmata* are sacred objects "designated particularly to be offerings" (Gernet, p. 128). In other words, they are often used within religious commerce, in particular, in cases of behavior meant to exemplify lavish generosity. A first example is the tripod of the seven sages: "It is a matter of a reward to be given to the most wise person, given to each of seven whose names were transmitted in varying from throughout classical antiquity. This reward is at times a tripod, at times a cup or goblet made of gold. Most often it is Thales who wins it first, and who then passes it on to someone whom he acknowledges as being wiser than he; this recipient in turn gives it to a third, and so on until the object passes from the seventh wise man's hands back into those of Thales" (Gernet, p. 131). Traditionally, the circular transmission of the sacred object is associated with a banquet. Gernet notes that "in the end, as if the value had increased as a result of the circulation, the tripod is consecrated to a god" — Apollo — (Gernet, p. 132). In this legend we find once again an association between concepts of value, circularity, monetary symbols, banquets and the sacred.

An analysis provided by Daniel de Coppet of funeral cycles among the 'Aré'Aré may be even more telling because it is based directly on ethnographic material. De Coppet provides complex descriptions of traditional funeral feasts given in the honor of an individual who has died a natural death (de Coppet, 1968). In these feasts, three groups are brought together: *O*, the group of officiants, all men closely related to the deceased party; *G*, the group of gravediggers, and *I*, those individuals who come to offer money on the day of the ceremony. The monetary exchange that takes place on this occasion is very intense, the money being used consisting of pearls and/or dolphin teeth. De Coppet distinguishes between four kinds of monetary prestations. The first, which is of particular interest in our context, is a matter of money given to members of group *I* to those of *G*. These prestations are called 'ascending' ones because the money is placed on a platform on which the officiants stand. Receipt of this money involves an obligation in return, for the money must be given back to the *G* by the *I*, but only two or three years later, on the occasion of the 'return funeral feast'. Thus an economic cycle is connected to a religious one. During the two or three years that separate the beginning and the end of the cycle, the gravediggers have full use of the money they received. Generally they share it with those who helped them to prepare the funeral ceremony, in particular those who provided pigs and taro puddings.

But the first funeral cycle contributes to the development of a new one to the extent that the officiants, whose deceased was honored, are now obliged to offer a small sum of money when one of the members of *I* has the occasion to serve as either the gravedigger or officiant in a funeral ceremony. This group of social relations results in a network that is "so tangled that it seems difficult to describe it satisfactorily. It expresses a general imbrication of all individuals within exchange relations that are at once frequent, countless, and of small amounts. All of the money given away on the occasion of a funeral feast will be regained in any number of ulterior ceremonies given in the honor of the dead" (de Coppet, 1970b, p. 33). Here we encounter once more the image of a social totality having monetary circularity as its basis.

Why is the circle never broken? How is the obligation to put the money back into circulation expressed? What prevents private appropriations of some of the money? These are very complex questions. We have the beginning of a response to them when we observe that the monetary sign already belongs to someone else, namely, to God. This kind of belief is illustrated by Max Weber, who points out that in the Protestant ethic, money is a good that belongs to God. This is why it should not be dissipated in useless individual expenditures. The Protestant asceticism leads to a prodigious accumulation of capital.

Weber notes that in such a conception, man is only a manager of the goods that have been committed to him by the grace of God, and it would be scandalous to dissipate these goods for the sake of personal pleasure. Like the servant in the parable, the holder of these goods must account for every penny that has been placed in his keeping.

It also seems that the tripod of the seven wise men is a sacred object from the beginning. According to one version of the myth, it was created by Hephastes; in another, it is found at the bottom of the sea. This tripod is said to have been a present at a divine wedding. Later it was owned by Helen of Troy, who threw it into the ocean when an oracle told her to do so (Gernet, p. 136). It would seem that in archaic societies, the circularity that is the foundation of the community is conceived of by means of a postulation of a transcendental referent.

De Coppet's analysis of the 'Aré'Aré ceremonial system confirms this hypothesis and renders it more precise. Let us note first of all, in keeping with our critique of economism, that in the 'Aré'Aré society ordinary everyday exchanges of goods for money are infrequent and have little importance. Yet monetary exchange within ceremonies is frequent. Foodstuffs and the products of handicraft are exchanged quite adequately by means of barter, whereas *ceremonial forms of exchange always have a monetary connotation.* Monetary prestations, and in particular those described above, have the result of constituting what de Coppet calls the image of the ancestors. The sum of 'ascending money' gathered during a given ceremony is proclaimed aloud at the end of the day. It indicates the precise value of the deceased party, the size of his influence and renown outside the group of his immediate friends and family. At the same time, the sum measures the solidity of the bonds uniting the diverse groups who take part in the feast. Yet the 'ascending money' gathered together in one funeral feast will be dispersed until another feast, in a regular and incessant movement that de Coppet calls 'continuous time'. The latter is the ultimate expression of global social cohesion: "One can say that there is no higher authority than this circulation of ancestral images. In the form of money, the ancestral value is tangible, it is given in movements that govern everything, and in particular, all men" (Barraud *et al.*, p. 459). In this example, the exteriority that founds monetary exteriority is the world of ancestors. *Such is the source of monetary legitimacy.* Money expressed the power of the dead, a power that is ambivalent, being held to be at once maleficent and beneficent. By virtue of its supernatural power, money must be handled with a great deal of care. Before making a gift of it at a feast, it is necessary to ask the ancestors to make it harmless. Private ownership has no place within such a system. *We have shown that hoarding is closely linked to individual values, and it is highly significant that hoarding is prohibited in 'Aré'Aré society*: "Hoarding is a crime that can lead to the sickness and death of the guilty party. In order for money not to harm the person who holds it, money must circulate endlessly, moving from one person to the next . . . otherwise, its force causes death" (de Coppet, 1970b, pp. 35—36). Here we encounter a kind of obligation analogous to those analysed by Marcel Mauss in relation to the *hau* of the Maoris. These are symbolic universes where the subject/object dichotomy is not always relevant: "The thing received is not inert. Even when given up by the donator, it still has something of him in it. By means of this thing, the donator has a hold on the receiver . . . the *hau* pursues anyone who holds it . . . it wants to return to its place of birth, to the sanctuary of the forest of the clan and owner" (Mauss, pp. 159—160). In the 'Aré'Aré, this lack of a clear and universal distinction between subject and object is expressed in the belief that there is no ultimate boundary between the world of the living and the world of the dead: "Exchanges between the living can only be understood by taking into account the relations between the living and the dead which is the primordial manifestation of the global order of the society and implicitly of the whole universe" (Barraud *et al.*, 1984, p. 510). This expression of the global values of a society in its exchanges are what make the latter a 'total social phenomenon'. This dependence on the dead, who are the source of monetary value, takes an exemplary form in the case of an old person having no heirs: "The end of a genealogical line leads to . . . the sudden interruption of the cult of ancestors; money has no way to exist and loses its *raison d'être*" (de Coppet, 1970b, p. 37). Thus we witness the death of money. It goes back to where it came from. The old man gathers up all the money, opens his father's grave and consigns this treasure to it forever.

There are quantitative differences between the various archaic conceptions of social exteriority, but there are no qualitative breaks. In each instance it is a matter of a belief in a form of radical alterity or absolute transcendence. All that varies is the specific way in which this transcendence is expressed. The passage from the fragmentary and plural image of the world of the ancestors, as in the case of the 'Aré'Aré, to the world of monotheism amounts to a progressive concentration of the sacred. We are tempted to say that the intensity of the traces of individualism within holist societies is a function of the degree to which the sacred is concentrated. These traces can be seen in cases where the circulation of money is broken, and also in any other case where there appears some more or less explicit form of private appropriation. In such cases modern money begins to emerge. But in order to develop these remarks, we must first set forth the sacrificial hypothesis.

### 2.2. *Sacrificial Money*

The general hypothesis that money has a religious origin can best be developed by bringing together the work of Bernhard Laum and René Girard. Within such a perspective, money and the sacred are seen to be the simultaneous products of a single mechanism, namely, sacrifice. In this manner, the close link between these two social realities is unambiguously revealed. Girard's theory of the sacred explains the production of the sacred as a form of absolute transcendence by means of a mimetic polarization of individual rivalries around a single object, the scapegoat or sacrificial victim. Ritual sacrifice is a repetitive recreation of this sensation of emotional unity that gathers individuals together in their unanimous relation to the sacred Other. This theory takes account of the essential features that we have used to characterize the holist form of the sacred — absolute exteriority and radical alterity. The theory also makes it possible to develop the analysis begun by Laum.

According to Laum, if in Homer's time a bull served as a unit of value, this was because the animal had a sacrificial role. Hocart defends a similar thesis: "In fact, according to my hypothesis the most ancient form of monetary operation is no other than the gift made to the priest in payment for services rendered during the sacrifice" (Hocart, p. 111). In the Girardian perspective, the entire social and religious system is organised around sacrifice. The object sacrificed — the bull — has in the course of historical evolution been substituted for the original victims, who were human beings. As Desmonde notes in his defense of Laum's ideas, "it seems likely that the numerous rituals of the bull god in Crete originated in the practice of human sacrifice" (Desmonde, p. 103). The sacrificial bull is divided into portions that are then shared among the celebrants. The sacrificial meal thus would appear to be the forerunner of coinage: "The share in the bull's flesh at the public meal also constituted a legal means of payment. The word nomos, which in later times signified the sacred law of the community, originally meant 'distribution' and denoted the division of the sacrificial offering. Similarly, the term dais, which was applied to the holy meal, was also the word for 'distribution'. . . . The laws of the state cult stipulated precisely the manner of apportionment to the gods, to the presiding authorities, and to the citizens. . . . The portion of the body of the victim functioned as an economic payment made legitimate through the justice of the state in ritual distribution. The portion which each communicant received expressed his worth, the degree of his esteem in the group" (Desmonde, pp. 115–116). The evolution of ritual progressively leads to the substitution of symbolic objects, such as pieces of metal that have been struck as coins, for animals. The Girardian theory fully accounts for this phenomenon since it recognizes that the very essence of sacrifice is the replacement of a generalized violence in the group with a unanimous violence directed against the emissary or surrogate victim. The very nature of the process of victimization and sacrifice is to polarize the attention of the group on surrogate objects that provide the focal point of mimetic unanimity.

Laum's hypothesis makes it possible to take the many dimensions of primitive money into account, such as the role of priests and temples in the circulation of money or the symbolisms that accompany protomonetary objects. We see that money designates a transaction made with a divinity, just as our earlier analyses suggested. An important issue was that of linking up money's function as a unit of value and the system of estimating the value of offerings. We see that what emerges is a price system founded on the relative values that diverse goods have in relation to sacrificial ritual. This theoretical perspective will now allow us to take up the issue of the private appropriation of money. The legend of the ring of Polycrates is most instructive in this regard.

"Polycrates was a tyrant during the second half of the 7th century B.C. As his untarnished happiness had inspired the jealousy of the gods, he was told by an oracle to give up some of his wealth, and more precisely, whatever was of greatest value to him. He was thereby led to throw the famous ring into the sea. But contrary to every expectation, the ring was washed up on shore, discovered, and brought back to the tyrant. Unable to obey the gods' command, Polycrates is punished: only his total poverty will expiate a prosperity that had lasted too long" (Gernet, p. 112). Sacrificial themes appear very clearly in this myth, which Gernet uses in his discussion of the origin of value in the *algamata*. The ambivalent nature of wealth is brought fully to the fore.

As in the situations we have already mentioned, the holding of precious objects is dangerous. In fact these objects belong to a divinity, either as a past offering or as a potential one. In either case, the object is part of a sacrificial ritual. To possess it is to give rise to the jealousy of the gods. It seems only natural to us to perceive behind this transcendental jealousy a more earthly and concrete form of jealousy — the jealousy felt by all of the members of the community when they witness the spectacle of a disproportionate individual fortune. In the collective mentality, this excessive good fortune seems like a kind of embezzlement of the gods' funds — or in other words, of what belongs to the community as a whole. Thus Girard's theory reveals the concrete bases of the obligation to return the ceremonial gift: social envy, which could potentially degenerate into a generalized form of violence (Girard, 1985). But at the same time the *agalma* has a beneficent side as well. As an object that emerged from sacrificial polarization, it can serve as a substitute victim and thereby contains violence. From this perspective, the *agalma* is a talisman that protects its owner. Whence the value of such precious objects. They provide a kind of insurance against collective violence (one thinks in this regard of the many ancient evil eye amulets known as gorgonia, adorned with the face of the Gorgon Medusa). Money's function as an amulet is based on a sacrificial ritual that made the mimetic unanimity converge upon a single point. The destruction of money can calm violence as well, for in religious thought, this destruction is not understood as a negation of value, but as its return to the realm of the gods. This thesis concerning the conflictual origins of value is clearly illustrated in the tale of Polycrates. The fact that he fails to return the wealth reveals the social action behind the story, the group's collective condemnation of Polycrates himself. Having failed to divert the violence onto the object, he is obliged to pay in person. The story ends with him in total poverty, but it is easy to imagine that what is really at stake is his life.

This same structure reappears in the 'Aré'Aré society in relation to the figure known as the peace-master. This figure is, very approximatively, to Melanesian society what the tyrant was to the ancient Greek city-states, namely, someone distinguished by his fame. The culminating point in the career of the great man is a feast called the peace-master's feast. It begins in the same manner as a funeral ceremony, for the same prestations of 'ascending money' are made. Various people place sums of money on the top of a platform (Barraud *et al.*, pp. 455—459). The sum collected during the feast measures the size of the peace-master's fame. But something changes at this point: "Everyone mimics the foibles of others, people exchange the grossest insults, and the most lewd and scandalous copulations are simulated. In the evening . . . everyone on the way home destroys the gardens and coconut palms of the peace-master, who does not even have the right to complain. These acts of destruction . . . set the date of the next big feast back by several years. Rare are the peace-masters who can afford to offer more than two of them, so costly is this form of glory, and so complete is the ruin it brings" (Barraud, *et al.*, p. 456). Here we find the same schema as in the tale of Polycrates: an offensive level of wealth, the jealousy of the community, and the same final ruin. But in the case of 'Aré'Aré, the story is not only told in a symbolic form where the jealousy, the return of the ring, and the final condemnation are caused by the gods. In the peace-master's feast, the real actor in such dramas is fully present: an angry and jealous crowd. As a result, the logical links in the action are made clear. In the beginning there is an attempt to usurp the legitimate power, which belongs to the ancestors. This attempt is based on the use of large amounts of 'ascending money'. The peace-master organizes a ceremony that is a kind of parody of the funeral ritual that consecrates the dead (one thinks of the end of the feast of Trimalchio in *The Satyricon*). When he announces the amount of money present on the platform, the peace-master institutes himself as an ancestor; in front of everyone he claims a sacred status for himself. As one may well imagine, this situation leads to a social crisis of no small magnitude. The typical symptoms of the sacrificial crisis, the crisis that leads to the immolation of the victim in Girard's theory, are apparent in the description of the strange scenes that take place on the evening of the feast (Girard, 1972, Ch. 5). This violence ends up being exorcized in the total destruction of the goods owned by the peace-master.

Thus the 'Aré'Aré ceremony makes it possible to understand the Greek myth, for it makes explicit two fundamental points that the tale of Polycrates only hints at. First of all, the appropriation of sacred wealth, and more specifically, of monetary goods, must be seen as an attempt to destabilize the dominant form of social legitimation. This point leads us to underscore the fact that Polycrates is a tyrant. Perhaps there is a close link between the rise of coinage and tyranny. The numismatic phenomena could be closely related to the crisis of the Greek city-states during the 7th and 6th centuries B.C. (Servet, pp. 112–122; Shell, 1979, pp. 11 and p. 62). This politico-religious hypothesis is supported by Herodotus's tale of the tyrant Gyges, who usurped power in Lydia by killing the legitimate king, Candaules. It seems to be quite plausible that it was during this tyrant's reign that coinage first appeared. On the other hand, the 'Aré'Aré ceremony clearly identifies the role of the violent crowd within the sacrificial logic. The crowd is no longer hidden behind the vague and indeterminate mask of divine jealousy.

The comparison of these two structures nonetheless reveals one quantitative difference. In 'Aré'Aré society, the emotional intensity is almost entirely concerned with the maleficent side of hoarding. Within Greek civilization, a more positive connotation is given to monetary power when associated with such social figures as sacred or magic kingship and tyranny. This difference expresses the growing importance of individualist values within classical antiquity, particularly in the form of the conquest of power by the tyrants. Melanesian culture, on the other hand, reduces the individualist components to a minimum, so much so that it is integrated in the very moment that it is manifested. This explains why in Melanesia, no matter how striking the peace-master's success may be, he is nonetheless "overwhelmed by the order of values that he serves with the greatest fidelity" (Barraud *et al.*, p. 518).

## CONCLUSION

Modern economic theory tends to describe money as a convention, in a sense that is close to the definition proposed by David Lewis. The circulation of commodities is perceived as a series of buyings and sellings oriented towards the consumption of goods, in relation to which money is a neutral instrument. That money is conventional is in no way a problem because the nature of the means that makes the circulation of commodities possible does not matter. "Its functional existence absorbs its material existence", as Marx puts it. In such a world — a world fully described by the quantitative theory — the demands for money and for commodities are established in exactly the same way, for in both cases it is a matter of individuals trying to maximize their personal utilities. The stability of various demand functions is the precise corollary of the stability of preferences. Hoarding disturbs this perfect order. In the hands of the greedy hoarder, money becomes the source of a particular power, commonly known as 'the power of money' – the power to stop the circulation of commodities and to change the rules in one's own favor. What is expressed in hoarding is essentially a critical perspective on the representations that guide the actions of the other agents on the market. Hoarding expresses a doubt about the correctness of the conventions that govern the system. Concentrated in this figure is the modern, individualist dimension of money, its role as the instrument used to express critical rationality.

Let us underscore the analogy between the figure of the hoarder and the figure that Louis Dumont perceives as standing at the beginning of individualism, namely the 'individual-outside-the-world', an individual who is distant from the social world and who subjects its value to a process of 'relativisation' (Dumont, p. 36). For Dumont, this attitude can arise in societies that are dominated by the holist ideology if the 'individual-outside-the-world' is also an 'individual-in-relation-to-God', for example, a Brahman. This privileged relation to divinity is what gives the individual a status that allows him or her to criticize worldly values. These reflections lead us to an analysis of the idea that the power of the individual develops as a result of an intensification of religion, a process that culminates in protestantism. In a similar vein, Marcel Gauchet writes that "The greater are the gods, the greater their power, the more they are directly responsible for the invention of the world, the more will mankind be able, through them, to understand the origin. This is the fundamental paradox of the history of religion . . . the incremental increase of the gods' power has not been to the detriment of man's power, enhancing their subjection, but has, on the contrary, given men more power" (Gauchet, p. 23). The 'individual-in-relation-to-God' presents us with a paradoxical configuration, one identical to the one encountered in our analysis of hoarding: money authorizes individuals to express their distrust of the dominant economic convention, while drawing its power from the very conventions that it defies.

Our cursory examination of the ethnographic documents confirms our analysis. Money as a means of storing value, as a source of arbitrary power in relation to the rest of society, and as a break in the infinite chain of reciprocal bonds, is a threat to the holist order, which is why it is the object of so many prohibitions. What should be stressed here is that this use of money corresponds to a very basic change in its status: money is no longer a means but an end, for in hoarding, money as such is what is desired. This astonishing mutation itself gives rise to transformations that alter the space of values. Money levels objects, effacing their qualitative differences. In money the things that are most remote from each other find a common point and enter into contact. By progressively dissolving all particular meanings and values, money reveals the interchangeable nature of desire. With money, a new perspective on the world is created, a perspective in which all ends are equivalent and possibilities are indeterminate. The open-ended future stands against the circular conception of time that closed the horizons of holist societies. Within these societies, "exchanges are first and always a necessary ordering of the past. If they at the same time involve an opening upon the future, this future will be a repetition of the past. To think on the other hand of exchanges as a series of dealings with the future or as productive of a changing process is our modern manner of apprehending them" (Barraud *et al.*, p. 431). The unpredictable nature of the future thus appears to be characteristic of a modern institution of temporality. Money is no longer a mere means, but is sought after for its own sake because its possession makes it possible to have a different perspective on the world, the experience of the relativity of desire. Money is the incarnation of relativity of all things. Again we see the link between this type of behavior and modern rationality which "by eroding the old frameworks and requiring neutral, homogeneous legitimations of beliefs, pushes the world into becoming a *Bundle of Hypotheses*, and thus a home fit for instrumental rationality, if not perhaps for much else" (Gellner, p. 81). Money and critical reason seem to go hand in hand.

In this essay we have stressed the hypothesis that the desire for money for its own sake cannot exist in the absence of a certain type of collective belief that goes beyond the simple belief that social rules must be conventional. I have called this additional belief 'legitimacy', and it is a matter of cognitive operations that ground the truth of this desire. These operations put into play a number of symbolic analogies thanks to which institutions go beyond the stage of fragile conventions, as Mary Douglas puts it. Legitimacy is manifested in the emergence of collective representations by means of which the collectivity becomes aware of itself, and this emergence is governed by specular processes. Following this conception, the group's self-organization is not transparent. Such a hypothesis contradicts the individualist utopia and affirms that there is a dimension of society that necessarily stands outside individual awareness. As in Hayek's analyses, the market order causes a kind of holist knowledge to appear. Money is what stands as the obstacle to the individualist project of a fully contractual society. We can thus interpret the transformations of the monetary economy as being moved by the impossible desire for a complete dematerialization of money. This is the very same desire that is expressed and formalized in economic theory, as in the case of the 'quantitative theory of money'.

*CREA, Ecole Polytechnique, Paris*

## REFERENCES

Aglietta, Michel, and Orléan, André: 1982, *La violence de la monnaie*. Paris: P.U.F.

Barraud, Claude, de Coppet, Daniel, Iteanu, André, and Jamous, Raymond: 1984, 'Des relations et des morts: quatre sociétés vues sous l'angle des échanges', In: *Différences, valeurs, hiérarchie*. Paris: Editions de l'Ecole des Hautes Etudes, pp. 421—520.

Barre, Raymond: 1970, *Economie politique*. 6th Ed. Vol. 2. Paris: P.U.F.

Bouv resse, René (1981), *Karl Popper*, J. Vrin, Paris.

Coppet, Daniel de: 1968, 'Pour une étude des échanges cérémoniels en Mélanésie', *L'Homme*, **4**, 45—57.

Coppet, Daniel de: 1970a, 'Cycles de meurtres et cycles funéraires, esquisse de deux structures d'échange', In *Echanges et communications, mélanges offerts à Claude Lévi-Strauss pour son soixantième anniversaire*. The Hague: Mouton, pp. 559—582.

Coppet, Daniel de: 1970b, '1, 4, 8; 9, 7. La monnaie, présence des morts et mesure du temps', *L'Homme*, **10**, 17—39.

Desmonde, William H.: 1962, *Magic, Myth, and Money*. New York: Free Press.

Douglas, Mary: 1986, *How Institutions Think*. Syracuse, N.Y.: Syracuse University Press.

Dumont, Louis: 1983, *Essais sur l'individualisme*. Paris: Seuil.

Depuy, Jean-Pierre: 1986, 'L'autonomie du social', *Cahiers du CREA*, **10**, 229—273.

Depuy, Jean-Pierre: 1989, 'Self-Reference in Literature', *Poetics*, **18**, 491—515.

Durkheim, Emile: 1978, *De la division du travail social*. Paris: P.U.F.

Ellis, Howard E.: 1934, *German Monetary Theory, 1905—33*. Cambridge, Mass.: Harvard University Press.

Ewald, François: 1986, *L'Etat-Providence*. Paris: Grasset.

Feyerabend, Paul: 1975, *Against Method*. London: Verso.

Frankel, S. Herbert: 1977, *Two Philosophies of Money: The Conflict of Trust and Authority*, New York: St. Martin's.

Gauchet, Marcel: 1985, *Le désenchantement du monde*. Paris: Gallimard.

Gellner, Ernest: 1985, 'The Gaffe-avoiding Animal or A Bundle of Hypotheses', In *Relativism and the Social Sciences*. Cambridge: Cambridge University Press, pp. 68—82.

Gernet, Louis: 1982, *Anthropologie de la Grèce antique*. Paris: Flammarion.

Girard, René: 1972, *La violence et le sacré*. Paris: Grasset.

Girard, René: 1985, *La route antique des hommes pervers*. Paris: Grasset.
Goux, Jean-Joseph: 1984, *Les monnayeurs du langage*. Paris: Galilée.
Haesler, A. J.: 1986, 'Logique économique et déterminisme social', *Bulletin du Mauss*, **18**, 105—137.
Hahn, Frank: 1984, *Equilibrium and Macroeconomics*. London: Basil Blackwell.
Hayek, Friedrich A. von: 1973, *Rules and Order*. Vol. 1 of *Law, Legislation and Liberty*. London: Routledge & Kegan Paul.
Hayek, Friedrich A. von: 1976a, *The Mirage of Social Justice*. Vol. 2 of *Law, Legislation and Liberty*. London: Routledge & Kegan Paul.
Hayek, Friedrich A. von: 1976b, *Denationalisation of Money*. London: The Institute of Economic Affairs.
Hicks, John R.: 1981, *Valeur et capital*. Paris: Dunod.
Hocart, Arthur M.: 1973, *Le mythe sorcier*. Paris: Payot.
Kaldor, Lord Nicholas: 1939, 'Speculation and Economic Activity', *Review of Economic Studies*, **7**, 1—27.
Keynes, John Maynard: 1937, 'The General Theory of Employment', In *Monetary Theory*. R. W. Clower (Ed.), Penguin Education, pp. 209—223, 1973.
Keynes, John Maynard: 1960, *A Treatise on Money*. Vol. 2. London: Macmillan.
Keynes, John Maynard: 1973, *The General Theory of Employment, Interest and Money. Collected Writings*. Vol. 7. Cambridge: Cambridge University Press.
Kindleberger, Charles P.: 1978, *Manias, Panics, and Crashes*. New York: Basic Books.
Knapp, Georg F.: 1905, *Staatliche Theorie des Geldes*. Leipzig: Duncker & Humblot.
Knight, Frank H.: 1921, *Risk, Uncertainty, and Profit*. Boston and New York: Houghton Mifflin.
Kuhn, Thomas S.: 1970, *The Structure of Scientific Revolutions*. 2nd Ed. Chicago: University of Chicago.
Laidler, David and Rowe, Nicholas: 1980, 'Georg Simmel's Philosophy of Money: A Review Article for Economists', *Journal of Economic Literature*, **18**, 318—387.
Laum, Bernhard: 1924, *Heiliges Geld*. Tübingen: Verlag von J.C.B. Mohr.
Lepage, Henri: 1985, *Pourquoi la propriété?* Paris: Hachette.
Lewis, David K.: 1969, *Convention: A Philosophical Study*. Cambridge, Mass.: Harvard University Press.
Lucas, Robert E.: 1981, *Studies in Business Cycle Theory*. Cambridge, Mass.: M.I.T. Press.
Marx, Karl: 1883, *Das Kapital: Kritik der politischen Ökonomie*. 3rd Edition. Hamburg: Otto Meissner. Trans. Samuel Moore and Edward Aveling. *Capital: A Critique of Political Economy*. New York: Modern Library, 1936.
Mauss, Marcel: 1983, *Sociologie et anthropologie*. 8th Ed. Paris: P.U.F.
Orléan, André: 1989a, 'Pour une approche cognitive des conventions économiques', *Revue Economique*, **40**, 241—272.
Orléan, André: 1989b, 'Mimetic Contagion and Speculative Bubbles', *Theory and Decision*, **27**, 63—92.
Polanyi, Karl: 1944, *The Great Transformation: The Political and Economic Origins of Our Time*. Boston: Beacon.
Popper, Sir Karl: 1956, *The Open Universe: An Argument for Indeterminism*. W. W. Bartley, III (Ed.). Totowa, New Jersey: Rowman and Littlefield.

Samuelson, Paul A.: 1968, 'What classical and neo-classical monetary theory really was', In *Monetary Theory*. R. W. Clower (Ed.). Penguin Education, 1973.

Samuelson, Paul A.: 1976, *Economics*. 10th Ed. New York: McGraw-Hill.

Sargent, Thomas J. and Wallace, Neil: 1976, 'Rational Expectations and the Theory of Economic Policy', *Journal of Monetary Economics*, **2**, 169—183.

Sargent, Thomas J. and Wallace, Neil: 1982, 'The Real-Bills Doctrine versus the Quantity Theory: A Reconsideration', *Journal of Political Economy*, **90**, 1212—1236.

Servet, Jean-Michel: 1984, *Nomismata*. Lyon: Presses Universitaires de Lyon.

Shackle, George L.S.: 1972, *Epistemics and Economics*. Cambridge: Cambridge University Press.

Shell, Marc: 1979, *The Economy of Literature*. Baltimore and London: The Johns Hopkins University Press.

Shell, Marc: 1982, *Money, Language, and Thought*. Berkeley: University of California Press.

Simmel, Georg, 1982. *The Philosophy of Money*. Trans. Tom Bottomore and David Frisby. London and Boston: Routledge & Kegan Paul.

Vroey, Michel de: 1987, 'La possibilité d'une économie décentralisée', *Revue Economique*, **38**, 773—806.

Wallace, Neil: 1980, 'The Overlapping Generations Model of Fiat Money', In *Models of Monetary Economies*. J. H. Kareken and Neil Wallace (Ed.). Minneapolis: Federal Reserve Bank of Minneapolis, pp. 49—82.

Weber, Max: 1958, *The Protestant Ethic and the Spirit of Capitalism*. Trans. Talcott Parsons. New York: Scribner.

JEAN-JOSEPH GOUX

# PRIMITIVE MONEY, MODERN MONEY

I am very interested in the distinction made by Orléan between two origins: the origin of modern money which seems now purely conventional, unconvertible, and the origin of primitive money. It seems to me that this division between two origins can solve many problems, but that this duality of origins can also raise new interesting but difficult issues. And the questions I have in mind after the close reading of the text are related to this double origin, or more precisely, to the relationship between the two regimes of money described in the text, not only the historical problem of the *passage* from one to another, but also the structural problem of persistence (or not) of the primitive status into the modern status. To sum up roughly: the primitive money would be *sacrificial*, the modern money would be *non-sacrificial* (no transcendence, no alterity, but self-institution). The problem would be: what kind of *historical*, or rather *critical* consequences should be drawn concerning this difference? Is the non-sacrificial status of modern money a stable possibility or a catastrophic limit? Is the modern man doomed to live this new status as a *loss* (*lack* of transcendence, alterity, sociality) or as a conquest? If the primitive money is sacrificial, is there a sacrificial dimension *hidden* under the modern (or post-modern) money, with the result that this irreducible dimension threatens to rise up again and to re-structure violently the system? Or is modern money the historical symptom of the profound resolution crisis, a kind of interiorization-resolution which permits this self-functioning?

Now I want to stress two main points:

1. What struck me in the anthropological material exploited by Orléan, is the symbolic position of archaic money which seems to confirm the structural homologies I tried to demonstrate by other ways (Cf. 'Numismatics' in *Symbolic Economies*). Money and the *father*. Money and the *phallus*. Money and the *language*. It is not too difficult to discover these *isotopia* in the Orléan analysis.

(a) Money expresses the power of the dead, the ancestors, the fathers. It is the 'symbolic' or the 'dead' father. This relationship between



*Francisco J. Varela and Jean-Pierre Dupuy (eds), Understanding Origins*, 145—149.

money and fathers is certainly very primitive but, nevertheless, one can notice an interesting vestige of it in our contemporary society: on the American one-dollar bank note, the image of a founding father (Washington) can be seen, with a mention of God ("In God we trust").

(b) Money is also, according to Gernet's analysis, the *agalma*, circulating treasure. The phallic signification of the Greek *agalma* was underlined explicitly in one of Lacan's text. (Cf. 'Subversion of the subject and dialectic desire').

(c) Money is also, according to de Coppet's description, a means to open or to close the communication, a sacred language.

So there is a clear symbolical and structural solidarity between money, language, father, phallus. One can say that in an archaic society, where the purely economical level of social exchange is not clearly drawn, there is an undifferentiated notion of 'value', 'power', 'efficacy'. 'Money' can embody this undifferentiated value with parental, sexual, linguistic dimensions. In our societies where various differentiated levels of 'value' prevail, we have to reconstruct theoretically the system of symbolical homologies which is hidden, partly unconscious.

But what I want to notice is the fact that, concerning all these elements, it is possible to find a fundamental moment of exclusion or separation. The symbolic father is the *dead* father, that is to say, the father existing in another world. The symbolic phallus is the phallus after the structural moment of *castration*. The genesis of money implies a radical separation between goods and money, the *exclusion* of the monetary element from the world of use-value, consumption, immediate enjoyment (*jouissance*). The language also is the result of a *separation*, the mourning of the thing itself. When I say 'a flower' it is "the absentee of the bouquet", as Mallarmé states.

So, it seems that the shaping of a *standard of value* or meaning supposes a radical exclusion, and a kind of sacrifice. I do not want to go further in this analysis, but it was worth underlining this striking homology. There is no *unit, measure, standard of evaluation*, without a genetical and structural stage of exclusion of the standard which is also the separation between *one* element becoming sacred and the other profane elements.

2. I have a second point to make. I want to stress the importance of the

*Greek moment* to understand the problem of the passage from sacrificial to non-sacrificial money.

I recall that the Greeks are supposed to be the inventors of money, that is to say, circulating metallic money coined and stamped by the State and functioning as a general equivalent of all other goods.

They are also the initiators of a philosophy which concerns self-reflexion, self-consciousness.

They invented also policy (and not only politics), that is to say, they invented and discussed political constitutions, new modes of institution, including even a fictitious institution, namely *Utopia*. Castoriadis, in articles about the Greek *polis*, shows the importance of this self-institution and he underlines the vocabulary which stresses it: *autonomos, autodikos, autotéles*.

So we find in the Greek moment a very coherent and striking picture of the germ of modernity: the beginning of modern money, the beginning of the movement of self-institution of society, and the beginning of individualism.

Now, it can be said that the Socratic precept "Know thyself" formulates both the individualistic and self-referential logic of this moment. How to understand the ethical revolution corresponding to this moment? Girard's approach can certainly enlighten this Socratic moment, because it is very precisely the surpassing of sacrificial logic which is a logic of projection toward the other, of false imputation. Self-knowledge is the movement inward which allows us to break the imaginary imputation of the other, the projection of our negative side, our own shadow. In my view, Socrates's teaching corresponds precisely with this breaking of the projection. The core of the ethical Socratic teaching can be found in the following precept: it is better to suffer an injustice than to commit it. In other words: it is preferable to be a victim than a persecutor, and, at the limit: it is preferable to die, to be murdered, rather than to commit an injustice. That is the Socratic teaching against Calliclès. "I tell you, Calliclès, that to be boxed on the ears wrongfully is not the worst evil which can befall a man, nor to have my purse or my body cut open but that to smite and slay me and mine wrongfully is far more disgraceful and more evil. And to despoil and enslave and pillage, or any way at all to wrong me and mine is far more disgraceful and evil to the doer of the wrong than to me who am the sufferer" (*Gorgias*, p. 509). That is the fundamental ethical reversal.

It is very clear that this ethical position can be interpreted in

Girardian terms: to stand up for the victim and not for the stronger forces of the persecutor, to undermine the *pharmacos* logic.

Now, is it a mere historical coincidence if the Greeks invented circulating money and if Socrates, in the same cultural conjuncture, could promote this new ethical requirement in which the Just, the Righteous accepts the place of the victim?

Can't we see a parallelism between the passage from archaic sacrificial money to Greek modern money and the passage from *pharmacos*-oriented ethics to Socratic ethics?

In Socratic ethics, the *pharmacos* is no longer the Other. But each individual has to choose his possible own identification with the so-called *pharmacos* (beaten, tortured) to be righteous. In other terms: each individual has to make a *self-sacrifice*. So, I think that the general logic of the *self* is not the complete disappearance of the sacrificial dimension but a change in the orientation of the sacrifice: the self-sacrifice of our own criminal violence, by identification with the *pharmacos* and not with the persecutor. So the Socratic self-knowledge is coherent with the self-sacrificial aspect of his ethics.

It is clear that there is a parallel between this ethical position and the truth-seeking, or namely the scientific thinking initiated by Greek philosophers. Concern for the victim, the breaking of the *pharmacos*-oriented attitude can be correlated with the scientific attitude or objectivity. The problem would be to know what new attitude comes first. But, any way, the objective attitude implies a movement of de-projection: the withdrawal of our own feeling attributed to things or to persons. The imaginary accusation, the false charge against a victim, the projection of our own negative feeling on the *pharmacos* is perhaps the archetype, or, at least, the very significant case, of the mechanism of projection. Socrates breaks this mechanism when he sees the *pharmacos* as the Just, the Righteous, and when he exposes the ugliness of the agression.

Now, the notion of personal *reciprocity* and personal *equality* (and not clanic or tribal reciprocity) supposed by monetary exchange has an obvious link with the Socratic statement. Let's recall that Aristotle deals with money in a chapter of his *Ethics.* Money is an instrument of *equalization*, not distributive justice, not rectificatory justice, but *commutative* justice, grounded on reciprocity.

In the Greek historical passage from a dominant distributory and rectificatory justice to a dominant commutative justice, there is a shift from money as standard of value, sacred and transcendental archetype of evaluation, to money as medium of exchange. In Egyptian society, for example, there already was a fixed standard of value deposited in the sanctuary, but no universal medium of exchange, no circulating money. The Greek innovation is the *circulation of this standard itself* as medium of exchange, general equivalent which takes part in the individual market, comes down in it. There is a de-transcendentalization of the standard. There no longer is one sacred object only, one sacred standard deposited in the sanctuary to measure the value of goods engaged in the barter, but a *circulating* universal equivalent. Each individual has a personal relationship with the standard, because the standard is not only the transcendental mean of measure, but, also, the means of circulation on the profane market. This new coincidence of these two functions seems the exact economic parallel of the individualistic transcendence of the new ethic. The structural moment of separation, exclusion, is no more embodied in the sacred and unique exclusion of the standard, but it passes through the subject himself. The alterity is inward. And each subject is the center of a possible universal evaluation.

From this point of view, Greek money is the germ of our modern money as described by Orléan. The movement of nominalization and self-reference leads toward present money. But to reach this moment, a new operation is necessary: money has to become a mere sign. In metallic money (silver, gold) which is a 'commodity money', the pure symbolicity of money as means of exchange is not completely apparent, only virtual. I am not sure that it is possible to completely ignore the historical and polical specificity and scope of the step from the 'commodity money' to the sign-money and unconvertible money. Especially the role of the State in monetary regulations becomes more powerful in the case of unconvertible monetary signs, and the socio-symbolic problem of representation is not the same. There is a split between things and signs. Signs seem to refer to other signs, in an indefinite play of referring, and not directly to things. It is a new symbolic device, different from the 'commodity money'-oriented economy.

*Rice University*

# PART III

# EVOLUTION AND THE DIVERSITY OF LIFE

STUART A. KAUFFMAN

# ORIGINS OF ORDER IN EVOLUTION: SELF-ORGANIZATION AND SELECTION

## I. INTRODUCTION

This article is written as a prolegomena, both to a research program, and a forthcoming book discussing the same issues in greater detail (Kauffman, 1991). The suspicion that evolutionary theory needs broadening is widespread. To accomplish this, however, will not be easy. The new framework I shall discuss here grows out of the realization that complex systems of many kinds exhibit high spontaneous order. This implies that such order is available to evolution and selective forces for further molding. But it also implies, quite profoundly, that the spontaneous order in such systems may enable, guide and *limit* selection. Therefore, the spontaneous order in complex systems implies that selection may not be the sole source of order in organisms, and that we must invent a new theory of evolution which encompasses the marriage of selection and self-organization.

The overall themes I wish to explore are these:

(1) Complex systems, such as the genomic regulatory networks underlying ontogeny, exhibit powerful 'self-organized' structural and dynamical properties.
(2) The kinds of order which arises spontaneously in such systems is strikingly similar to the order found in organisms.
(3) This raises the plausible possibility that the spontaneous order found in such complex systems accounts for some or much of the order found in organisms.
(4) The existence of strongly self-ordered properties in complex systems implies that selection must be acting on systems with their own 'inherent' properties; hence at a minimum, what is ultimately selected may often reflect a compromise between selection and the spontaneous properties of the class of systems upon which selection is acting.
(5) Such compromises reflect the fact that selection is in general a combinatorial optimization process in a rugged fitness landscape



*Francisco J. Varela and Jean-Pierre Dupuy (eds), Understanding Origins*, 153—181.

with many peaks, ridges and valleys. The typical structure of such landscapes and population flow upon them under the drives of mutation and selection, assure that attaining and maintaining high optima, with properties which are very rare in the class of systems under selection, usually cannot occur.

(6) The typical failure of selection to be able to *avoid* the typical properties of the class of systems under selection is therefore expected to sustain points (3) and (4) above. The spontaneous properties of complex systems under selection will often be similar to those generic in the class of systems under selection, not *because* of selection, but *despite* it.

(7) In turn, if many features of organisms reflect the generic properties of an entire class of systems under selection, not the particular successes of selection, then non-reductionistic theories based on analysis of those generic properties can be expected to be predictive of features of organisms. We should, in short, be able to predict many features found in organisms without needing to know the details.

(8) The capacity of selection to achieve rare highly functioning forms is governed by the statistical features of the adaptive landscape over which selection tries to pull an adapting population. As we shall see, many landscapes are 'bad' in the sense that adaptation is likely to become trapped on mediocre local optima which become ever more mediocre as the complexity of the entities under selection increases. We must envision yet a further broadening of evolutionary theory to encompass the possibility that selection has achieved entities with the internal properties which allow them to adapt on 'good' adaptive landscapes with high optima. Beginning conditions for this to occur will be discussed as well.

In preliminary summary: complex systems are self-ordered, promising relief to selection as the sole source of order in biology. The relief must be purchased, however, by a broadened theory which considers the capacity of such systems to adapt on rugged fitness landscapes, and the capacity of selection to alter the kinds of entities which exist, hence the kinds of landscapes they evolve upon. Evidently, we must take Darwin's central idea seriously, but move beyond it.

This article is organized as follows. In the second section I introduce the concept of an adaptive landscape, and characterize briefly the properties of adaptive walks to local optima in rugged landscapes. In the third section, I introduce a class of models which generate a family of tunably rugged landscapes and seek generic statistical features of adaptive walks on such landscapes. In the fourth section I discuss briefly the emergence of spontaneous order in models of genomic regulatory systems and the parallels between the generic properties of this class of systems and features of ontogeny. In the fifth section, I report briefly on the capacity of selection to act on genomic regulatory systems, and the limitations due to the statistical features of their rugged landscapes. In the sixth section I return to the issue of whether selection may achieve entities which have the internal properties allowing them to adapt on 'good' fitness landscapes, and discuss the extent to which this may circumvent or modify the general theme that the selection cannot avoid the typical properties of the class of systems upon which it operates.

## II. RUGGED ADAPTIVE LANDSCAPES

The adaptive landscape metaphor in evolutionary biology is at least as old as Wright (1932). The metaphor is limited in certain respects. Thus, it is well known in population genetics that a population undergoing selection on two or more loci, with two or more alleles per locus, may not flow 'up' the fitness landscape due to recombinational constraints between the two loci. Density and frequency dependent effects may also limit the simplest vision of a fitness landscape, where fitness is a function of genotype and attendant phenotype (Ewens, 1979). More generally, the fitness of an organism is a function of its own phenotype and a number of others of the same or other phenotypes, hence not a pure function of one phenotype alone.

Despite these and other limitations, the fitness landscape image is powerful, basic, and a proper starting point to think about selection. We conceive, next, of a very simple space of objects, namely peptides, and analyse the character of adaptive walks via 1-mutant fitter variants to local or global optima for a defined functional property of the peptides (Eigen, 1985; Smith, 1970).

### *Sequence Space*

Consider the set of all peptides of length 10. With 20 amino acid types, there are $20^{10}$ possible peptides with 10 amino acids. Define the 1-mutant neighbors of a peptide to be all those sequences which can be obtained by changing one amino acid to one of the 19 other possibilities. Thus, each peptide has $19N = 190$ 1-mutant neighbors. We are then conceiving of a peptide sequence space with all $20^{10}$ possible peptides of length 10, ordered in a high dimensional sequence space in which each peptide is a point, and is connected by a line to its 190 1-mutant neighbors. This space, in short, is a sequence space. Distance between points along connected lines passing through other points correctly represents the minimum number of changes in amino acids to convert one sequence to another.

We next need to define a fitness landscape. We do so based on the capacity of each peptide to perform some function. For example, we might measure the affinity with which each peptide binds a specific hormonal receptor on a cell surface. This measured affinity then can be thought of as a measure of the 'fitness' of each peptide with respect to that function. Because each peptide has a measured affinity, and the peptides are arranged as points in an ordered sequence space, the measured affinities constitute a fitness landscape over the sequence space.

Consider next the simplest possible caricature of an adaptive walk in peptide space. We imagine that the adaptive process begins with some arbitrary peptide. Next we imagine that at each generation one or more mutant variants of that peptide are produced. In the simplest case, a single mutant variant is produced, and it is mutant in a single amino acid. This 1-mutant neighbor may be fitter than the initial peptide. If so, let the adaptive process step to this improved variant. If not, let the process remain 'at' the initial peptide and try another 1-mutant variant. (I show below that this simplest case corresponds to a well defined reasonable population genetic situation.)

In this simplest image, an adaptive *walk* begins at an arbitrary peptide, and on each trial samples a 1-mutant variant of the current peptide, and 'moves' to that variant only if fitter. Thus, the process is constrained to pass via 1-mutant variants until it reaches a peptide which is fitter than all its 1-mutant neighbors.

Obviously, the character of an adaptive walk depends upon the disposition of the fitness values in the fitness landscape. That distribution might range from very smooth, such that neighboring peptides have highly similar fitness values, to rugged landscapes with many peaks ridges and valleys but still substantial correlation between the fitness values of 1-mutant neighbors, to fully uncorrelated landscapes in which the fitness values of 1-mutant neighbors were completely unrelated to each other.

Given any fitness landscape, the immediate natural questions which arise are: (1) How many improvement steps are taken on an adaptive walk before arresting on a local optimum? (2) How many local optima exist? (3) How many alternative local optima are accessible via branching adaptive walks from an initial arbitrary peptide? (4) How does that alter with the fitness of the initial peptide? (5) How 'fit' are the local optima with respect to the mean fitness in the space of peptides? (6) How do these properties depend upon the ruggedness of the landscape and the complexity of the entities (here peptides) under selection?

Given any fitness landscape, the actual flow of a population across it depends not only upon its structure, but upon population size, mutation rate, initial dispersion across the landscape, and other factors (Ewens, 1979). Thus, even if we understand the structure of such landscapes, we still will not understand population flow upon them.

Yet deeper questions arise: What kinds of entities, peptides or otherwise, have given types of fitness landscapes and why? Are some landscapes easier to evolve upon than others? Can selection achieve entities which 'live on' good landscapes? I return to these questions below.

### *The Fully Uncorrelated Landscape*

Simon Levin and I have recently analysed the case of adaptive walks on fully uncorrelated landscapes (Kauffman and Levin, 1987). These are constructed by assigning each peptide a fitness value drawn at random from a fixed underlying fitness distribution. To be concrete, we may take that distribution to be uniform between 0.0 and 1.0. Since the adaptive process we are considering passes via 1-mutant fitter variants, the actual fitness values are not important, and can replaced by the rank order of the peptides, from worst to best.

We have established the following features of adaptive walks on uncorrelated landscapes:

(1) The number of local optima is very large. For peptides with 20 amino acids it is $20^N/(19N+1)$. That is, the expected number of local optima is just the number of entities in the space divided by 1 plus the number of 1-mutant neighbors to each entity.

(2) Walks to local optima via fitter 1-mutant variants are very short: $\log_2 (19N)$. That is, the number of fitter variants encountered on a walk is just the logarithm base two of the number of 1-mutant neighbors to any peptide in the space. This implies that, for peptides length 10, walk lengths are on the order of 7 or 8, while for proteins of length 100, walks are on the order of length 11.
(3) Only a small fraction of all local optima are accessible from a single initial peptide via branching adaptive walks.
(4) As fitness increases at each step along an adaptive walk, the number of fitter 1-mutant neighbors decreases by a half. Thus branching walks to alternative optima initially have many alternative routes upward, and these dwindle to single routes which send upward to local optima. Such branching tress, in short, are bushy at the base and dwindle to single lineages.

*Brief Justification*

I have idealized a walk as sampling 1-mutant neighbors one at a time, and constrained to pass via fitter neighbors. This idealization corresponds quite closely to a population with a low mutation rate, in which the rate of finding a fitter variant is very low, while the fitness differentials between improved variants and 'wild type' is moderate. In this case, on a slow time scale, the population uncovers a fitter mutant, and on a fast time scale fitter mutant sweeps through the population. Thus, roughly, the population 'hops' to a fitter 1-mutant neighbor. Gillespie (1974) has shown that this limiting case corresponds to a continuous time, discrete state Markov process, with the population passing from one state to a neighboring fitter state.

Obviously, in reality, population may harbor 1-mutant, 2-mutant, . . . , $J$-mutant variants at the same time. Actual population flow upon a landscape is more complex than this simplest case, which I have adopted as a means to analyse the structure of the landscape itself, rather than the flow upon it.

*Universal Features of Long Jump Adaptation*

Real populations may be able to 'jump' long distances in peptide spaces in single mutational events. For example, recombination is a mutational process which may substitute a large number of amino acids simultaneously into a peptide or protein. Such a mutation can be thought of as jumping long distances across peptide space.

Consider next any correlated but highly rugged landscape, such as the Alps. If a mutational process jumps *beyond* the correlation length of that landscape, then the fitness of the point reached is fully uncorrelated with the fitness of the point left. This corresponds to the fact that altitudes of points 50 kilometers from any point in the Alps are essentially uncorrelated with the altitude of the point left. Thus, long jump processes in correlated but rugged landscapes encounter an uncorrelated landscape.

Levin and I (*ibid.*), established several universal features of such adaptive processes:

(1) After each improved variant is found, the waiting time to find the next improved variant *doubles*. Thus, the rate of improvement slows rapidly. The consequence is that the expected number of improvement steps, $S$, after $G$ long jump trials, is:

$$S = \log_2 G$$

(2) As is the case described above, the number of alternative fitter long jump variants is halved on average at each improvement step, thus branching lineages are bushy at the base and the rate of branching dwindles as fitness increases.

### *A Complexity Catastrophe Limits Selection as Complexity Increases*

We have uncovered a fundamental and previously unexpected limitation of adaptation on rugged landscapes via long jumps. The same limitation applies to adaptation via 1 or few mutant variants on fully uncorrelated landscapes: As the *complexity* of the entities under selection increases, there is a marked tendency for the *attainable local optima* or the states attained after long jump walks of any fixed length to fall toward the mean fitness of the space of entities! In the current case, the complexity of the entities is just the length of the peptides. As $N$ increases, the number of local optima increases, and the number of 1-mutant or $J$-mutant variants of each peptide increases, but the actual fitness of local optima, or fitness attained after a fixed walk length, *decreases*. This constitutes a kind of *complexity catastrophe*.

This limitation is terribly important. We have already assumed in our idealization that selection is *always strong* enough to pull a population to any optimum and hold it there, an assumption which is generally false. Even with this assumption, this new limitation says that as entities under selection become more complex, then in the limits of uncorrelated landscapes and walks via neighboring fitter variants, or for rugged landscapes and long jump adaptational walks, the fitness achieved *falls*. More complex means more mediocre. This is a powerful limitation. We must ask whether there may be conditions which circumvent this limitation. Those conditions would be upon the character of the *adaptive landscape itself*. We must ask whether there are conditions such that entities under selection can become more complex, yet the adaptive peaks do not dwindle. The answer is 'yes', as we discuss in the next section.

## III. GENERAL EPISTATIC INTERACTIONS AND RUGGED LANDSCAPES

What properties of complex systems control the ruggedness of the fitness landscapes upon which they may evolve? Do peptides adapt on highly correlated, or very rugged landscapes? What of large proteins? What of the coupled genomic regulatory system where the activities of structural genes are regulated by a variety of control genes in complex circuitry? What of the morphology of an organism? In general, the answers are unknown. In fact, to my knowledge, the question of the relation between the character of entities under selection and the ruggedness of their adaptive landscapes has never been addressed at all.

To approach this issue, I here introduce a general model of 'epistatic' interactions among traits in an organism. This model is similar in spirit to spin glasses (Anderson, 1985), and as we shall see, appears to have wide applicability.

Consider an organism with $N$ 'traits'. For the moment let us restrict attention to the case where each trait is simply present or absent, denoted by a 1 or 0. With $N$ traits, there are therefore $2N$ possible combinations of traits which might be present or absent. I wish to consider the fitness contribution of each of the $N$ traits. In general, the fitness contribution of each trait may well depend upon the presence or absence of others of the $N$ traits. Such dependencies are called epistatic interactions in population genetic models where one thinks of the fitness contribution of the gene at one locus, which may depend not only on the allele of that gene, but depend epistatically on the alleles present at some other loci. Because I wish to study the effect of *richness* of epistatic interactions on the ruggedness of fitness landscapes, I shall assume that the fitness contribution of each trait depends upon itself and $K$ other traits among the $N$. Increasing or decreasing $K$ then alters how many traits affect the fitness contribution of each trait. In addition to the constraint that each trait's fitness contribution depends upon the presence or absence of $K$ other traits, I shall add one further constraint for the moment. I assume that if trait $i$ matters to trait $j$, then trait $j$ matters to trait $i$. Within these constraints, I shall assign the $K$ traits mattering to each trait, $i$, randomly.

The fitness contribution of each trait, $i$, thus depends upon the presence or absence of itself plus $K$ other traits. Therefore the fitness contribution of the $i$th trait must be specified for each of the $2^{(K+1)}$ combinations of presence or absence of the $K + 1$ traits which affect trait $i$. To do so, I shall draw a different random number from a uniform distribution between 0.0 and 1.0 and assign it to each of the $2^{(K+1)}$ combinations of traits. The same is done for all $N$ different traits. Therefore, for any particular combination of $N$ traits being present or absent, the fitness contribution of all $N$ traits is specified. The fitness of the entire combination of $N$ traits is defined as the sum of contributions of all traits, divided by $N$. This yields an overall fitness for each combination of traits which lies between 0.0 and 1.0.

The model I have defined is a random epistatic fitness model, tuned by $N$ and $K$. It is open to many interpretations. For example, the 'traits' might instead be considered amino acids in a peptide. In the present case we would conceive of only two types of amino acids. Given a spatial interpretation of the $N$ bits as amino acids, we might choose the $K$ which matter to each side from neighboring or from random sites in the peptide. Obviously the model can be generalized to 20 rather than 2 kinds of amino acids. Under the interpretation that the 'traits' are genes, each open to 2 or many alleles, the model is interpretable as a model of epistatic interactions of $K$ loci upon each site, with the $K$ distributed in some defined order on neighboring or random sites on the chromosome. This genetic interpretation corresponds to a haploid model with a single copy of each chromosome.

Given $N$, $K$, and the distribution of $K$ among the $N$, what do the resulting fitness landscapes look like? Return to the simple case where each 'trait' is present or absent. Then the space under consideration is the ordered $N$ dimensional Boolean hypercube, where each vertex or point is a particular combination of the presence or absence of each of the $N$ traits. Further, each point is a 1-mutant neighbor of $N$ other points, attained by changing one trait from present to absent, or *vice versa*. Each point has a well defined fitness. Therefore a well defined fitness landscape exists over the Boolean hypercube.

### $K = 0$ *Corresponds To A Fully Correlated Landscape With One Optimum*

Consider the case where $K = 0$. Here the fitness contribution of each 'trait' is independent of all other traits. Since the fitness of the entire system is just the sum of the fitnesses of the $N$ traits, this model is identical to a simple additive fitness population genetic model with $N$ genes, each with two alleles and no epistatic interactions among the genes. In the present case, each trait, by chance, makes a higher fitness contribution if present, or if absent. Thus a single global optimum exists in which each trait is in its more valuable state, present or absent. Note that any other combination of traits is less fit than the global optimum. Further, any such suboptimal combination of traits lies on a connected pathway via fitter variants leading to the single (global) optimum. Each trait in its less valuable state need merely be 'flipped' to its more valuable state. Note next that the fitness of 1-mutant neighbors will be nearly the same. This follows since flipping a single 'trait' will, on average change the total fitness less than $1/N$. That is, the $K = 0$ landscape is a highly *correlated* landscape.

Deeper insight into the present $K = 0$ case is readily attained. Because each fitness contribution has, in the present case, been drawn at random from a uniform distribution between 0.0 and 1.0, and the fitter value is the fitter of two random draws, on average the less fit value is 0.333 and the more fit value is 0.666. In a system with $N$ 'traits', the expected fitness of the global optimum is 0.6666. Furthermore, the expected fitness of the global optimum is obviously *independent* of $N$. As $N$ increases, the fitness of the global optimum remains 0.6666. If an adaptive walk begins with an arbitrary combination of traits, its fitness will be about 0.5. Half the traits will be in their less valuable state, thus on the order of $N/2$ improvement steps occur on an adaptive walk to the single and global optimum. Along that walk, the number of fitter neighbors decreases by one on each step upward.

### $K = N$ *Corresponds To A Fully Uncorrelated Fitness Landscape*

At the opposite extreme, suppose the fitness contribution of each trait depends upon all traits. Then $K = N - 1$, or for large $N$, $K$ effectively equals $N$. Consider a particular combination of $N$ traits. Change one trait from present to absent. Then the context of all $N$ traits has changed. For each we shall have specified a fitness value at random from the uniform distribution. Therefore flipping a single trait yields a 1-mutant neighbor whose fitness is an entirely new sum of $N$ random samples from the uniform distribution. That is, the fitness values of 1-mutant neighbors in the space are entirely random with respect to one another. The $K = N$ case therefore corresponds to a fully uncorrelated fitness landscape.

We already understand the character of adaptive walks on such a landscape. There are a very large number of local optima. Walks to local optima take on the order of $\log_2 N$, rather than $N/2$ for the $K -$ 0 limit. The number of fitter neighbors is halved after each improvement step. From any initial combination of traits only a small fraction of the local optima are accessible.

Among the most critical features of $K = N$ landscapes is that they exhibit the *complexity catastrophe* described above. As the number of traits, $N$, increases, the fitness of attainable local optima *decreases* toward 0.5, the mean of the space.

### $0 < K < N$. *Tunably Rugged Fitness Landscapes*

Clearly, increasing $K$ from 0 to $N$ corresponding to increasing the epistatic interactions among the traits in the system, increases the ruggedness of landscapes from fully correlated to fully uncorrelated. The intent of the model now becomes clear, for the parameters $N$, $K$, and the distribution of $K$ among the $N$, generate a *family* of expected landscapes with statistically characterizable adaptive walks. It may well be the case that this family covers a vast spectrum of 'real' adaptive landscapes which arise in a variety of combinatorial optimization problems from the Traveling Salesman problem (Lin and Kernighan, 1973), to walks in protein spaces during maturation of the immune response (Kauffman and Levin, 1987). At a minimum, this family of landscapes should offer beginning intuitions about the requirements on system construction and the ruggedness of the resulting landscapes.

To approach this issue, my colleague E. Weisberger and I have written a general computer program. In Tables I and II I show the results of numerical Monte Carlo trials for different values of $N$ and $K$, and for two different distribution rules on the $K$. In the first, we conceived of circular peptides (to avoid end effects) and required the $K$ sites bearing on each site to be its $K/2$ neighbors on either side. In the second set, Table II, we removed the restriction that $K$ be reflexive, and allowed each site to be affected by $K$ randomly chosen other sites. The results do not appear to depend upon this difference strongly. Thus, $N$ and $K$ determine the major features of the class of landscapes.

The numerical results confirm the general expectations derived above:

(1) Both Tables confirm that for $K = 0$, the single global optimum has a fitness of 0.66666, and that the optimum is independent of $N$.

TABLE I

Neighboring $K$. Upper values indicate fitness of optima; lower values the number of steps to attain the optimum.

| $K \backslash N$ | 8 | 16 | 24 | 48 | 96 |
|---|---|---|---|---|---|
| 0 | 0.646 | | 0.656 | 0.659 | 0.665 |
| | 5.5 | | 13.59 | 25.31 | 49.77 |
| 2 | 0.698 | | 0.699 | 0.703 | 0.705 |
| | 5.14 | | 12.22 | 23.49 | 46.20 |
| 4 | 0.6983 | | 0.696 | 0.7011 | 0.699 |
| | 4.58 | | 10.35 | 20.27 | 38.31 |
| 8 | 0.659 | 0.679 | 0.680 | 0.688 | 0.682 |
| | 3.67 | 5.73 | 8.66 | 16.29 | 28.70 |
| 16 | | 0.646 | 0.655 | 0.660 | 0.658 |
| | | 4.33 | 5.63 | 11.02 | 28.06 |
| 24 | | | 0.628 | 0.641 | 0.643 |
| | | | 4.48 | 8.65 | 15.75 |
| 48 | | | | 0.598 | 0.613 |
| | | | | 5.02 | 9.90 |
| 96 | | | | | 0.566 |
| | | | | | 6.10 |

TABLE II

Random $K$. Upper values indicate fitness of optima; lower values the number of steps to attain the optimum.

| $K \backslash N$ | 8 | 16 | 24 | 48 | 96 |
|---|---|---|---|---|---|
| 0 | 0.649 | 0.662 | 0.657 | 0.659 | 0.664 |
| | 5.6 | 7.54 | 12.67 | 25.31 | 48.61 |
| 2 | 0.676 | 0.679 | 0.687 | 0.690 | 0.691 |
| | 5.48 | 9.15 | 13.68 | 25.60 | 49.11 |
| 4 | 0.690 | 0.700 | 0.712 | 0.714 | 0.717 |
| | 4.79 | 8.62 | 13.17 | 24.23 | 47.94 |
| 8 | 0.676 | 0.693 | 0.702 | 0.712 | 0.715 |
| | 3.62 | 6.64 | 9.91 | 19.55 | 37.57 |
| 16 | | 0.644 | 0.653 | 0.669 | 0.686 |
| | | 4.20 | 5.90 | 11.67 | 24.83 |
| 24 | | | 4.50 | 9.14 | 17.61 |

Both Tables confirm the expectation for the $K = N$ case that as $N$ increases, the fitness of local optima *decreases* towards 0.5. The complexity catastrophe on uncorrelated landscapes is real.

(2) For $K = 0$, walk lengths are on the order of $N/2$, and the number of fitter neighbors declines by 1 per step. For $K = N$, walk lengths are on the order of $\log_2 N$, and the number of fitter neighbors declines roughly exponentially.

But for values of $K$ between 0 and $N$, new features emerge.

(1) Note for $K$ small and fixed, for example, $K = 2$, 4 or 8, that as $N$ increases, the fitness of optima actually increases towards an apparent asymptote which becomes independent of $N$, and which is *larger* than the global optimum for $K = 0$. That is, a small amount of epistatic interaction actually *helps create a landscape with optima of higher fitness!* And the fitness appears to become independent of $N$ for large $N$. Thus, for $K$ fixed (and perhaps small) complexity the sense of increasing $N$ does not lead to a complexity catastrophe. The space retains optima of high fitness.

(2) What happens if $K$ increases proportionally to $N$? Both tables

show that for $K = N/2$, $K = N/3$, and $K = N/4$, as $N$ increases the fitness of optima first increases above the $K = 0$ case, then decline below it, and presumably fall towards 0.5 as $N$ grows sufficiently large. Thus, if $K$ is proportional to $N$, it appears that a complexity catastrophe occurs. Increasing complexity leads to landscapes with lower optima. $K$ proportional to $N$ would mean, intuitively, that as the number of traits in an organism increases, or number of amino acids in a peptide, that a constant fraction of those trails affect the fitness contribution (i.e. function) of each trait or amino acid.

These last two observations carry the following implication. Consider $K$ as a function of $N$. For $K$ constant as $N$ increases, the complexity catastrophe is averted. For $K$ proportional to $N$, it is not averted. Thus for $K$ equal to some monotonically increasing function of $N$ a transition must occur between landscapes which do not and do exhibit the complexity catastrophe. What that boundary is remains unknown.

How generalizable are these results? In the present simplest model we have sampled fitness values from a uniform distribution between 0.0 and 1.0. We do not yet know, but strongly conjecture that the qualitative results are independent of the underlying distribution, which might be uniform, Gaussian, exponential, etc. so long as the fitness contribution for each combination of 'traits' is sampled from the *same* distribution. Indeed the results may generalize to a process which samples from a set of different distributions.

To summarize tentative conclusions: If as $N$ grows large the number of 'traits' which impinge epistatically on any trait is bounded below some constant or slowly increasing value, then such systems adapt on rugged landscapes which avert the complexity catastrophe. Beyond some critical rate of increase in $K$, the complexity catastrophe sets in. I suspect that the implication of this general argument holds quite generally. For complex entities to evolve, the number of parts which directly impinge upon any part probably must grow very much more slowly than the number of parts in the whole. I turn in the next section to a description the self-organized dynamical behavior of model genomic regulatory systems. There it turns out that genetic networks in which each gene is directly affected by only a few other genes exhibit marked order reminiscent of real ontogeny. In the subsequent section I present evidence that genomic systems of low connectivity also adapt on correlated landscapes which avert the complexity catastrophe. Thus, the same property, low $K$, appears to abet orderly behavior, and capacity to evolve well!

## IV. SELF-ORGANIZED BEHAVIOR IN GENETIC NETS OF LOW CONNECTIVITY

Metazoans such as mammals have on the order of 100 000 different structural and regulatory genes. Since Jacob and Monod discovered that products of one gene could activate or repress the activities of other genes, biologists have come to think of the genomic system as a kind of biochemical computer. The structural and regulatory genes are linked into some kind of circuitry, regulating and coordinating one another's behavior. It is a cannon of current developmental biology that all the diverse cell types of an organism contain the same set of genes coordinated by the same regulatory circuitry. While there are minor exceptions to this cannon, it is close enough to truth for our current purposes. In terms of this cannon, the fundamental problem of cell differentiation is: Why are cells different from one another, despite harboring identical genomic systems? The answer, of course, is that cells differ because different subsets of genes are expressed in different cell types. It follows that it is the genomic circuitry and regulatory system which coordinates and engenders these different patterns of gene expression.

The presumptive complexity of a genomic system with about 100 000 genes regulating one another's activities bodes serious epistemological problems. Molecular genetics is fast uncovering the molecular machinery by which one gene acts on a second gene. Structural genes, which code for proteins, typically are flanked by specific DNA sequences which are bound by the protein products of other genes. This binding can turn on or turn off transcription of the structural gene to messenger RNA, thence to protein. Such adjacent DNA sequences, because they regulate adjacent genes on the same chromosome, are called *cis* acting regulatory genes. The upstream genes whose products bind to cis acting sequences are themselves called *trans* acting, because they can exert regulatory influences on genes located on different chromosomes. Regulation is more complex than mere control of transcription. Different points in maturation of messenger RNA, and translation to protein are also subject to regulation. For our purposes I shall lump all these into the term 'genomic regulatory network'.

The genomic regulatory network is subject to evolution in two quite different ways. First, of course, point and recombinational mutations may alter DNA sequences leading to new useful structural genes and proteins. Some of these altered proteins may themselves be *trans*-acting proteins, hence their alteration may alter the effect of the regulating gene upon its target genes. But beyond such mutations, chromosomal mutations literally shuffle genes from one to another position in the set of chromosomes. Such mutations include duplications, and various processes which disperse duplicated genes to new locations such as translocations, transpositions, inversions, conversions. Consider the obvious consequences. If gene $A$ is adjacent to a *cis*-acting sequence $X$, and a transposition moves a new gene $B$ into position between $X$ and $A$, then $X$ may now control $B$ as well as $A$. If $B$ happens to be a *trans*-acting gene regulating a downstream cascade, then $X$ now regulates the same cascade via $B$. In short, chromosomal mutants literally scramble the circuitry in the genomic regulatory system.

From the foregoing, it follows that *evolution of new cell types* involves both the evolution of new genes, and new circuitry among old and perhaps new genes. Thus our intellectual task is formidable. We want to understand both how the genomic system in an organism underwrites its ordered ontogeny, and we want to understand how such regulatory systems evolve.

### *An Ensemble Approach*

It may be the case that to understand the genomic system in Man, or any other organism, its detailed structure, component by component, will require elucidation. Indeed, our intuitions based on complex systems such as current computer programs suggests this might be the case. Current computer programs are fragile. If two instructions are exchanged, typically the program is drastically altered. Nevertheless, I strongly believe that this intuition misleads us with respect to understanding genomic regulatory systems and their capacity to evolve. Instead, as I shall soon point out, complex systems of interacting genes can exhibit spontaneous order.

Two of the most readily accessible features of genomic regulatory systems are the numbers of genes which typically directly act on and regulate a single gene, and the kind of regulated responses which occur.

With regard to the first issue, genes in bacteria and viruses typically are directly regulated by 0 to perhaps 5 or 6 other genes and products (Kauffman, 1969, 1974, 1984). That is the *connectivity* of real genomic regulatory systems in these organisms is *low*. It now appears, but remains to be shown in more detail, that this is true in higher eukaryotes as well. What of the regulated behavior of a gene? For the moment, let us idealize the behavior of a gene and think of it as either active, 1, or inactive, 0. Then the behavior of any gene with respect to its regulatory inputs is given by a Boolean switching rule. For example, a gene might be active at the next moment if either its first, or its second or both inputs were active the moment before. This is the OR function. Alternatively, a gene might be active only if its first input were active and the second inactive the moment before. This is the NOT IF function. Within the idealization to 1 and 0 values, examination of bacterial, viral and eukaryotic genes shows that different genes are governed by different Boolean functions.

This leads to the following question. Suppose we consider models of genomic regulatory systems in which each gene is either active or inactive, hence a binary switching variable. If *all* we know about such networks is that, on average, any gene is directly regulated by only a few other genes, can we say anything about the expected structure and behavior of large genomic networks? Otherwise stated, does the *local* feature of *low connectivity* imply anything about the *global* behavior of such genomic regulatory systems? If so, that local feature would predict those global features, hence might explain them without need for detailed reductionistic analysis.

The form of this question is an *ensemble question*. That is, there is an enormous number of genomic regulatory systems with $N$ binary genes, each regulated by $K$ other genes. This set of systems constitutes an ensemble of all systems which might be built given $N$ and $K$. Thus, what we want to know is the typical or *generic* properties of members of this ensemble. To answer this question by numerical simulation, we should then construct model genomic systems entirely at random within the constraints that there be $N$ genes, each regulated by $K$ other genes. Specifically, this means deciding at random which among the $N$ genes are the $K$ inputs to each gene, then deciding at random for each gene which among the $2^{2^K}$ possible Boolean functions of $K$ inputs govern its behavior. That is, both the wiring diagram and the logic are chosen at random. Once chosen, the structure is fixed.

The results of many numerical simulations for different values of $N$

and $K$ have shown that for small $K$, for example $K = 2$, such networks exhibit powerful order which is strongly reminiscent of real cells. Briefly, such a network is a finite automaton. Each combination of activities of the $N$ genes is a state. Thus there are $2^N$ states. At each moment, the net is in a state. The genes each examine the activities of their inputs, consult their switching rule, then all genes synchronously assume their designated new activity value. Thus the net passes from a state to a state. Over time the system passes through a succession of states. Since there is a finite number of states, eventually the system reenters a state previously encountered. Thereafter, since the net is deterministic, it cycles repeatedly through this reentrant cycle of states. It is critical that many states may lie on sequences of states which flow to the same state cycle. But the network may harbor more than one state cycle. If so, some states flow to one of these assymptotic attractors, while others flow to the other state cycles. Wherever the system is released, it will flow to one state cycle attractor. Thus ultimately, the system comes to cycle about one or another of its state cycles.

The order which emerges spontaneously when $K = 2$ includes the fact that the lengths of state cycles are small, on the order of $N^{1/2}$, the number of states cycles in a net's behavioral repertoire is also about $N^{1/2}$, each cycle is a stable to transient reversal of the activity of most genes one at a time, and if unstable will flow to only a few of the remaining state cycles.

I have for a number of years wanted to see in this spontaneous order a deep similarity to real ontogeny. I make a single interpretation: I identify a *cell type* with a *state cycle attractor* of a genomic network. That is, a cell type is a stable recurrent pattern of gene expression engendered by the genomic logic. Given this interpretation, the theory makes a number of predictions which are surprisingly accurate. Thus, the number of cell types in an organism should increase as a square root function of the number of genes! This is surprisingly close to what is observed across many phyla (Kauffman, 1969, 1974, 1984). If a cell type is a state cycle attractor, then differentiation is the passage from one attractor to another. These models imply that any cell type can only flow to a *few neighboring cell types*, and from thence to a few other cell types. But this implies that ontogeny should occur along *branching developmental pathways*. Indeed, all contemporary multicellular organisms develop along just such branching cell lineage pathways. Presumably this has remained true for the past 600 million years. I return to this below, for I shall want to ask whether this feature of organisms reflects selection, or such a deep property of genomic systems that it is *impervious* to selection.

This class of models predicts many other features of current organisms, such as the similarity in gene expression patterns in different cell types (about 90% to 95% overlap), the existence of a large core of genes active in all cell types, the typical cascading consequences of deleting a single gene, and so forth (Kauffman, *op. cit.*).

Without belaboring the detailed results, the main point to cull from all these studies is that even randomly constructed model genomic systems with few inputs per gene exhibit order which is strongly reminiscent of that seen in ontogeny. This implies that our intuition about complex systems is wrong. Order emerges spontaneously. From this it follows that this order may account for the origin and persistence of such order in organisms. The spontaneous order is at least a handmaiden to selection. But of course, selection operates continuously. This brings us to the central problem which I believe faces evolutionary theory. We need to broaden evolutionary theory to understand how selection can act upon, with and through systems which have their own strongly self organized properties. How is selection enabled, guided and limited by those properties? Note that no area of science has dealt with this problem. I turn to beginning steps in the next section.

## V. SELECTION AND ITS LIMITS FOR DESIRED CELL TYPES

Can selection, operating both on structural genes and the 'circuitry and logic' of a genomic regulatory system, mold such a regulatory system to achieve arbitrary 'good' cell types? Obviously, to broach this question requires first of all some model of a cell type, and the ways in which mutation and selection can modify cell types. The ensemble of genomic network models described above provides just such a framework.

To be concrete, I shall ignore evolution of new structural genes coding for new useful proteins with novel enzymatic, structural or other features. Rather, I shall focus on evolution of the circuitry and logic to alter the coordinated patterns of expression among a constant set of structural genes. Within that limitation, and in the framework of the ensemble theory developed above, evolution can proceed via mutations in the regulatory *connections* between genes, thus altering the 'wiring diagram' of the genomic circuitry; or evolution can proceed by altering the regulated behavior of a gene as its inputs alter activities. That is, mutations can affect the Boolean function characterizing the response of any regulated gene. To be concrete, chromosomal mutations may alter the wiring diagram, or such mutations and point mutations may alter the local rule. For example, a mutation which prevents a repressor protein from binding to its *cis*-acting site may render an adjacent regulated gene constitutively (constantly) active. The gene now realizes the 'Tautology' logical function, always active.

### *Evolution Explores an Ensemble of Regulatory Networks*

In the second and third sections of this article I discussed the structure of fitness landscapes, where the points in the space were conceived to be proteins, each a 1-mutant neighbor of those other proteins obtainable by altering one amino acid in the protein's primary sequence. The concept of rugged fitness landscapes is far broader, however. Consider a genetic network with $N$ genes, and $K$ inputs per gene. Each such network is a member of a vast ensemble of all networks constructable with $N$ and $K$ as constraints. Each network is a 1-mutant neighbor of all networks which can be attained by altering one regulatory connection, or one 'bit' in the Boolean function regulating one gene. Therefore, we can, again, define the concept of a high dimensional space, where each point is an entire genetic regulatory network, and its 1-mutant neighbors are all those genetic networks accessible by minimal changes of wiring diagram or logic. Thus, we can consider evolution as occurring across a space of genetic regulatory networks.

Any property of such networks might serve as a property upon which selection acts. To be concrete, let us define the fitness of a genetic regulatory system by how closely the patterns of gene activity occurring on one or another of its state cycles match to an arbitrary pattern of gene expression. That is, we shall define the 'fitness' of a genetic network of $N$ genes by how closely one of its cyclic attractors comes to having a pattern matching an arbitrary 'desired' or 'good' pattern of gene expression across the $N$ genes. For example, if the arbitrary pattern among the $N$ genes is (101010101010 . . .), then the fitness of a given network is given by the fraction of genes whose activities match this pattern on the best matching state of one of the state cycle attractors. This therefore assigns a fitness between 0.0 and 1.0 to each network, hence generates a fitness landscape over the space of genomic systems.

As before, this fitness landscape may be more or less rugged. Also as before, in the simple case of adaptive walks constrained to pass via fitter 1-mutant variants, we can ask how many steps occur on the way to local optima, the number of alternative local optima accessible, the similarities of those optima. Of course we are interested in how closely such optima match the arbitrarily 'good' pattern for which we are selecting. That is, can selection achieve *arbitrary good* cell types? And we are interested, as the complexity of the genomic system, $N$, increases, whether selection leads to less fit optima. Does the complexity catastrophe creep in?

I note in passing that these selection studies are a root form of learning in massively parallel processing systems. Thus, in the neurobiological context, models of distributed content addressable memories use attractors as internal memories of external events (Kauffman, 1986). Asking whether mutation and selection can achieve desired attractors is a form of asking whether mutation and selection can be used to evolve parallel processing networks to have desired 'memories' (Hopfield, 1982). It is a further point of interest that any such system, if placed in a varying environment, will spontaneously classify its environments into equivalence classes given by those environments which leave a given attractor unaltered. That is these kinds of model genomic systems, if placed in an environment, naturally classify the *same* environment in different ways according to which attractor the system attains, and classify different environments as the *same* if those environments leave a given attractor unaltered. Nonlinear systems with multiple attractors exhibit the basis of classification and a kind of cognition of their environments. Selective evolution for useful behaviors is thus a start towards evolving systems which classify their environments, recognize and act upon them. See, for other ideas in this area, Varela's contribution to this volume.

### *Numerical Studies of $K = 2$ and $K = 10$ Boolean Networks*

The general results of numerical simulations of such adaptive walks are these (Hopfield, *op. cit.*).

(1) For Boolean networks with $N = 50$ and $N = 100$, and $K = 2$ or $K = 10$ the first major conclusion is that selection never achieves networks with fitness 1.0. Rather, fitness increases from an initial 0.5 and typically arrests at a local optimum with values of about

0.65 or 0.7. This means that, in general, adaptation by 1-mutant neighbors becomes trapped on local optima far below the logically possible perfect network. This is not abetted significantly by allowing 2-mutant or 5-mutant variants to be sampled.

(2) The second general result is that the rate of finding fitter variants slows rapidly. This implies that, as fitness increases, the fraction of fitter neighbors dwindles rapidly.

(3) The third general result is that *long jump* adaptation, in which up to half the connections or one fourth the bits in Boolean functions are altered at once, reveals the expected general features of long jump adaptation on rugged landscapes. Thus, the waiting time to find fitter variants doubles on average after each improvement step. And as $N$ increases, the fitness attained after any fixed number of tries *falls* towards the mean fitness in the space of systems, 0.5.

(4) The fitness landscape for $K = 2$ is highly correlated. One mutant alterations to connections or logic make little difference to fitness. By contrast, the fitness landscape for $K = 10$ networks is far more rugged. Altering one connection, and even one bit in a Boolean function, typically alters a net's fitness drastically. Thus, $K = 2$ nets adapt on a more correlated landscape than do $K = 10$ networks.

(5) Given (4), it is interesting to ask if $K = 2$ networks exhibit the complexity catastrophe for adaption via 1 or 2 or 5 mutant neighbors. (They do for long jump adaptation, of course.) The answer is *no*. As $N$ increases, the fitness of local optima remains essentially constant. The results for $K = 10$ networks are not available.

Point 5 is particularly interesting. Recall from Section III that for that $NK$ spin-glass like model, we found that for $K$ small, optima were insensitive to $N$, while for $K = 0.5N$ or $K = N$, the complexity catastrophe sets in as $N$ increases. Thus there, and here, low connectivity yields a correlated landscape and the capacity to avoid the complexity catastrophe as complexity, $N$, increases. I stress, of course, that the two models are not at all identical. In Section III, the $NK$ models are taken as general epistatic models. In Section VI, the $NK$ models are *dynamical models* of Boolean switching networks with $M$ genes, and $K$ inputs per gene. Landscapes in the switching networks are defined by attractors. Nevertheless there may be deep homologies between them with respect to ruggedness of landscape.

A major conclusion of both *NK* models, then, is that the complexity catastrophe can be averted by systems having the proper properties to adapt on correlated landscapes where optima do not whither toward the mean as complexity increases. In both cases this seems to require low connectivity among a rich system of many interacting parts. I turn next to wondering if selection can select for systems which adapt on 'good' landscapes.

Two even more fundamental observations about these simulations must be stressed:

First, the fact that strong selection, that is selection which is always able to pull an adapting population 'to' a fitter variant, becomes trapped on local optima far from the logically possible Boolean network which matches a given pattern of gene expression precisely, very strongly implies that real selection acting on real organisms *cannot* sculpt arbitrary patterns of gene expression as the cell types of an organism. Whatever our intuitions have been, cell types almost certainly are not 'precise' in that sense. Evolution must make do with local optima, the best attainable on adaptive walks from whatever the starting point may be. The 'developmental program' beloved of developmental biologists is almost certainly not able to evolve to arbitrary logics from any given starting point.

Second, and perhaps most fundamentally, selection in rugged landscapes typically cannot *avoid* the typical features of members of that landscape. Adaptive walks in a space of genomic regulatory systems: (1) become arrested on local optima; (2) in the limit of long jump adaptation climb higher at an ever slower rate remaining far from global optima; or (3) in the face of a high mutation rate which disperses a population away from attained optima, wander ergodically among suboptimal genomic systems. In all cases, it appears, selection is limited in its capacity to attain genomic systems with features which are extremely untypical of the space of genomic systems in which adaptive evolution is occurring. But if true this bears the deepest consequences. For if selection can typically not *avoid* the typical or generic features of the genetic regulatory systems in the ensemble explored, then those typical features should 'shine through', and be found in organisms, not because of selection, but despite it. Here then is a major theme. If selection cannot avoid the generic features of complex genomic regulatory systems which are members of some definable ensemble, then those generic properties may prove to be biological universals. Their explanation would not lie in details of structure, nor details of selective forces in branching phylogenies, but in mere ensemble membership. As phase transitions in water are understood as a form of critical phenomena in a general class of systems without precise accounting for locations of water molecules at the transition, so some or many ordered properties of organisms may find their explanation in the generic properties of the ensemble of which the organisms are members.

I return, in this light, to the fact that model genomic systems with few inputs per gene have the property that each state cycle attractor, which models a cell type, is a 'neighbor' of only a *few* other model cell types. Thus, this broad class of genomic models predicts that any cell type can differentiate into only a few other cell types, and they in turn to a few others. As noted earlier, this implies that ontogeny must be based about *branching* cell differentiation pathways, as is indeed the case. Does this organization of ontogeny, presumably present since the late Precambrian, reflect an achievement of selection? Or is it a property which is so deeply embedded in the entire ensemble of genomic regulatory systems accessible to selection, so deep a property of parallel processing nonlinear dynamical systems, that selection cannot avoid this property? I suspect the latter. And add that it is no trivial property, for it is the same property which assures the homeostatic stability of cell types in the face of perturbations, which assures that a variety of specific and non specific 'inductive' agents induce the *same* differentiation transformation, hence which underlies the subsequent elaboration of the logic of development. The possibility that such branching pathways reflect self-organization, not selection, I suspect, is the harbinger of many possibly new universals reflecting the balance between self-organization and selection.

But all of this brings us to a yet harder issue. Is selection doomed to wander among systems in a given ensemble? Or can selection actually change the way genomic systems are constructed, hence change the ensemble being explored, its generic properties, and the character of fitness landscapes upon the ensemble? Can selection find ensembles with 'good landscapes'?

## VI. SELECTION FOR 'GOOD' LANDSCAPES

What kinds of systems have 'good' landscapes? It is clear that fully uncorrelated landscapes are not 'good' in at least two senses:

(1) First, in a correlated landscape, advantage may be taken of the correlation structure to 'look' where the looking is good. For example, many landscapes may prove to be self-similar. Spin-glass energy landscapes are an example (Anderson, 1985), and my first *NK* models in Section III may also be self-similar. That means that the landscape is fractal, with small hills and local optima located in the 'sides' of similarly shaped, but wider and higher hills and optima, which in turn are on the sides of still larger wider hills with higher optima. In such a correlated landscape, not only does it pay to look in the vicinity of modest local optima by jumping just far enough away to avoid trapping, but it may be the case the local optima carry global information about the general region of space with yet higher optima. Thus, one optimum is a good starting place to find yet higher optima. In contrast, in a fully uncorrelated landscape, locations of optima can carry no information about good regions of space. The search may as well be entirely random.
(2) In fully uncorrelated landscapes, the complexity catastrophe creeps in. As systems become more complex, the optima attained relapse towards the mean fitness in the ensemble being explored.

In contrast, we have so far encountered only one class of entities which adapt on landscapes which avoid these problems. Such entities appear to have the property than any component directly interacts with a small number of other components in the system, and that small number remains bounded small in some as yet unknown way as the number of components in the system increases. Can selection 'tune' *K* and keep it small?

Let us consider two different levels of systems upon which selection may act. Consider proteins first. Small peptides, that is sequences of amino acids with about 20 or fewer members, probably adapt on very uncorrelated landscapes, for substitution of any single amino acid seems likely to have marked effects on the behavior of the peptide. By contrast, in large proteins, many amino acid substitutions have little effect on the overall architecture and function of the protein. Thus, large proteins almost certainly evolve on more correlated landscapes than small peptides. Evolution has opted for large proteins, by and large. This may reflect the commonly held view that large proteins can fold better, hence form better enzymes and structural components of cells. But it may also reflect the fact that in selecting for large proteins, selection has also achieved the kinds of entities which adapt on more correlated landscapes hence can actually *adapt better.* Large proteins live on landscapes with connected walks to higher optima than peptides do. And probably the linear structure of proteins assures that the dominant interactions of any amino acid is with its few neighbors in the primary sequence. To a lesser extent, folding brings distant regions into contact, but still any amino acid interacts directly with only a few others. Thus, as the length of a protein becomes large, the number of strong interactions bearing on the contribution of any amino acid remains bounded and small. Proteins are thus probably well set up to adapt on correlated landscapes by their inherent nature.

What of genetic regulatory systems? What is low connectivity? Nothing other than the property that the one gene is directly regulated by only a few other genes or their products. But this is a restatement of *molecular specificity.* Enzymes and other liganding agents with high specificity will, in general, discriminate among a vast number of molecular variables and bind only a few close cognates. Selection for high specificity is also selection for low connectivity in a genomic regulatory system. And selection for low connectivity is simultaneously selection for systems which exhibit the globally ordered dynamics of homeostasis so reminiscent of actual cell types. In turn, it happens that selection for low connectivity and consequent homeostatic dynamical behavior also yields genomic regulatory networks which adapt on highly correlated landscapes which avoid the complexity catastrophe!

Two points warrant attention. First, by selection for large proteins, or high specificity, selection may be changing the adaptive landscapes of the kinds of entities which are evolving. Thus we do seem to confront the fact that selection can act on entities to achieve those with 'good' landscapes. This would not seem to require any group selection argument. This implies that we need to broaden evolution theory not only to include the balancing forces of selection operating on systems with self-organized properties, but further, to begin to understand how selection may operate to cull entities better able to adapt because they adapt on good landscapes.

Second, in both the case of selection of large proteins and genomic systems of low connectivity, the drive towards good primary function — proteins which are good enzymes and genomic systems with stable homeostatic dynamics — appears to bring with it almost gratuitiously selection for entities which happen to adapt on 'good' landscapes. Is this fortuitous? Or is there some deeper reason hinting that complex systems that 'work well' also 'adapt well'?

Quite obviously, if we take the Kantian stance and ask "What must organisms be such that they can evolve adaptively?", we become aware of how very much we have to learn. And if we look beyond evolution in biology to evolution in technological economies and society at large, we may well wonder whether there be found general principles of adaptation in complex systems.

## SUMMARY

I have tried to suggest that evolution is a complex combinatorial optimization process on some form of rugged fitness landscape.[1] This leads us to realize that adaptation on such landscapes tends to become trapped on local optima, or diffuse away from such optima if mutation rates are large enough with respect to selective forces. We were led to wonder if there might be quite general statistical features of rugged landscapes holding over many combinatorial optimization problems.

Adaptive evolution confronts the profound opportunities afforded by the self-organized properties of complex dynamical systems. I have here broached only genetic regulatory networks, but similar issues appear to arise in models of the origin of life (Eigen and Schuster, 1979; Kauffman, 1986; Dyson, 1985; Rossle, 1974), of connected metabolisms, (Kauffman, 1986), and morphogenesis.[2] This surely implies that we must build a larger theory which marries Darwin's idea of selection to the self-organized properties of the entities that selection was privileged to operate upon. Here we confront the possible limitations in the power of selection to avoid the typical properties of such systems, hence potential universals in biology, and an escape from the epistemological necessity of full reductionism. But beyond this, selection appears able to fashion the entities it operates on to be ones with useful landscapes. We thus need a still broader theory, encompassing this possibility, and then the probability that within such a 'selected' ensemble, selection will still not be able to avoid the typical properties of that selected ensemble, hence again an escape from necessary reductionism in our explanations.

Only the surface has been touched. We want a true theory of biological evolution embracing self-organization, selection, and historical accident.

*University of Pennsylvania, Philadelphia*

## NOTES

[1] A large literature on combinatorial optimization exists, but not often from the evolutionary perspective. In part this is based on simulated annealing (see, for instance, Kirkpatrick *et al.*, 1983).

[2] A very large number of articles and books are based on Turing's (1954) reaction-diffusion model of morphogenesis. Authors include Brian Goodwin, James Murray, Stuart Kauffman, Hans Meinhart, and Lionel Harrison over the past decade, writing in *Science, Nature, J. Theor. Biol., Developmental Biol.* Other articles by Lewis Wolpert, George Oster, Peter Bryant and Susan Bryant, among others, are scattered through the same journals.

## REFERENCES


Anderson, P. W.: 1987, 'Spin glass Hamiltonians: a bridge between biology, statistical mechanics and computer science', in: *Emerging Syntheses in Science*, D. Pines (Ed.), Addison-Wesley.

Dyson, F.: 1985, *Origins of Life*. London: Cambridge University Press.

Eigen, M.: 1987, 'Macromolecular evolution: dynamical ordering in sequence space', in: *Emerging Syntheses in Science*, D. Pines (Ed.), Addison-Wesley.

Eigen, M. and Schuster, P.: 1979, *The Hypercycle*. New York: Heidelberg: Springer Verlag.

Ewens, W. J.: 1979, *Mathematical Population Genetics*. New York: Heidelberg: Springer Verlag.

Gillespie, J. H.: 1974, *Theor. Population Biol.* **44**, 167.

Hopfield, J. J.: 1982, 'Neural networks and physical systems with emergent collective computational abilities', *Proc. Natl. Acad. Sci.* **79**, 254.

Kauffman, S.: 1969, 'Metabolic stability and epigenesis in randomly constructed genetic nets', *J. Theor. Biol.*, **22**, 437.

Kauffman, S.: 1974, 'The large scale structure and dynamics of gene control circuits: an ensemble approach', *J. Theor. Biol.* **44**, 167.

Kauffman, S.: 1984, 'Emergent properties in random complex automata', *Physica*, **10D**, 145.

Kauffman, S.: 1986, 'Autocatalytic sets of proteins', *J. Theor. Biol.*, **119**, 1.

Kauffman, S.: 1991, *Origins of Order and Self-organisation in Evolution*, New York: Oxford Univ. Press.

Kauffman, S. and Levin, S.: 1987, 'Towards a general theory of adaptation on rugged fitness landscapes', *J. Theor. Biol.* **128**, 11.

Kauffman, S. and Smith, R. G.: 1986, 'Adaptive automata based on Darwinian selection', in: Farmer, Lapedes, Packard and Wendroff (eds.), *Evolution, Games and Learning. Models for Adaptation in Machines and Nature*. Amsterdam: North Holland.

Kirkpatrick, S., Gelatt, C. D., Jr., and Vechi, M. P.: 1983, 'Optimization by simulated annealing', *Science*, **220**, 671.

Lin, S. and Kernighan, B. W.: 1973, 'An effective heuristic algorithm for the travelling salesman problem', *Oper. Res.*, **21**, 498.

Rossler, O.: 1974, 'Chemical automata in homogeneous and reaction diffusion kinetics', *Notes in Biomath.*, **B4**, 399.

Smith, John Maynard: 1970, 'Natural selection and the concept of a protein space', *Nature*, **225**, 563.

Wright, S.: 1932, 'The roles of mutation, inbreeding, crossbreeding, and selection in evolution', in: *Proc. 6th International Congress on Genetics*, Vol. I, p. 356.

JOHN DUPRÉ

# OPTIMIZATION IN QUESTION

Not very many years ago, what is now often referred to as the 'Panglossian Paradigm' reigned supreme in evolutionary biology. It was widely assumed, at least implicitly, that a central function of biology was to analyze organisms in such a way that the adaptive advantages, indeed the optimality, of their various features could be discerned. If it occurred to some imaginative biologist that snails would have been much better off running on caterpillar tracks, then some work would be needed to show why they were really better off dragging themselves around on a sheet of slime. For if they would have been fitter with the caterpillar tracks, then the omnipotent force of natural selection would surely have seen to it that they were so equipped. I exaggerate, but not all that much.

The most egregious and also most notorious examples of this program come from human sociobiology. At its not infrequent worst, this is a field in which anyone who can cobble together a selective scenario from a few bits of elementary game theory is thereby entitled to make portentous statements about the fundamental facts of human nature, and to do so with only the most cursory nod towards empirical facts about human behavior. In fact, this is really worse than the procedure parodied above. The correct analogy would have the biologist, after some impressive looking calculations in mathematical ergonomics, asserting that, contrary to appearances, snails really *do* run on caterpillar tracks.

What I have been irreverently describing are, of course, only the worst excesses of the Panglossian, or panselectionist, program. Beneath this tip of the iceberg there were many competent intelligent biologists doing useful empirical work, but work often guided by uncritical assumptions about the power of natural selection to produce optimal results. And I should certainly not imply that the debate is over, or that either the more respectable or the most egregious instances of Panglossianism do not continue to be produced. What has happened in the last decade is that the issue has come to the forefront of critical discussion. And many biologists have become quite deeply sceptical about the optimizing potential of natural selection*.



*Francisco J. Varela and Jean-Pierre Dupuy (eds), Understanding Origins,* 183—190.

I do not introduce this debate as a prelude to any kind of detailed discussion of the issues surrounding adaptationism and optimality. Rather what I want to point out as being an especially interesting and important feature of Kauffman's work, is its essentially post-Panglossian nature. The critique, if not rejection, of panselectionism, has raised fundamental questions in evolutionary biology, and Kauffman's work addresses some of the most important of these. First, and most obviously, if we cannot just assume that natural selection will produce an optimal result, how effective at increasing biological fitness is it? Kauffman's work on various kinds of adaptive landscape is a systematic attempt to address just this problem. And second, and perhaps even more significant, the assumption of panselectionism carried with it a dogmatic answer to questions about the origin of biological order: natural selection was quite sufficient for this task, and the suggestion that anything else might be necessary was generally associated with mysticism or, worse, special creationism. Attention to the limitations of natural selection has again raised the possibility that factors other than natural selection may have a central role to play in the explanation of natural order. Kauffman's investigations of the inherently self-ordering properties of genetic systems are a fascinating example of this move away from selectionist dogma.

I shall say a little more below about both the phylogenetic and the ontogenetic aspects of Kauffman's paper. But first, I would like to make a further point about how these two aspects are connected in relation to the critique of adaptationism mentioned at the beginning of these comments. Much, though by no means all, of the criticism of adaptationism has derived from the development of increasingly sophisticated views of the relation between genetic and phenotypic properties (phenotypic has now come to mean, effectively, non-genetic). The ideal selective scenario would be one in which a range of phenotypic features is exactly correlated with a set of alternative genes (technically: alleles at a particular locus). A particular animal, for example, might come with possible tail lengths of one inch, two inches, three inches, etc., depending on whether it had $allele_1$, $allele_2$, $allele_3$, etc. Selection determines that the animals with two inch tails are the fittest, and after a number of generations of selection, the entire population will consist of animals with $allele_2$, and consequent two inch tails. Unfortunately, the relation between genetics and development isn't even remotely like that. The way we divide up the features of an organism, and the way that selection distinguishes features of organism, typically does not appear to correlate with any simple or natural way of distinguishing its genetic features. Changes in a gene can affect numerous features of the phenotype, and many parts of the genetic material are relevant to the production of particular phenotypic features — not to mention, what Kauffman particularly focuses on, their effects on the activity of other genes. The upshot of all this is that evolutionary theory cannot just take development for granted, and assume that selection of fitter phenotypes will reliably select genotypes that will develop into similarly fitter phenotypes. On the contrary, it appears that the general character of the evolutionary process will depend on features, as yet very little understood, of the developmental program. My point, then, is that it is not just the desirability of exploring possible sources of biological order other than natural selection that makes Kauffman's ideas about ontogeny interesting, but that if we are to have any general theory of evolution it *must* incorporate some more adequate general understanding of ontogeny.

Let me now make a few more detailed comments about first the phylogenetic aspect of Kauffman's paper, the discussion of adaptive landscapes, and then the ontogenetic aspect. The appeal to adaptive landscapes in the early part of the paper (and elsewhere in Kauffman's work) is perhaps the part about which I am most inclined to scepticism. In general, the use of models has come to be recognized as a central aspect of scientific methodology, particularly in those sciences that deal with highly complex phenomena, such as all of the biological and social sciences. One reason for this is just the fact of complexity. Biological phenomena are typically much too complex for there to be any possibility of complete description (if, indeed, the notion of 'complete description' even makes sense in this context). In addition, they are almost always highly variable from instance to instance. Thus any progress in understanding must involve abstracting away from many possibly relevant features, and from much of the irrelevant variation. The employment of abstract models is the normal and unavoidable vehicle of this abstraction in biology. Biological models aim to distinguish a few fundamental features of a type of phenomenon, and if the features are well chosen, important processes can be isolated from the 'noise' of less relevant features. And again, the noise provided by irrelevant variation can also be filtered out by appropriate abstraction.

While I do not see that there is any alternative approach to dealing 185c typical biological phenomena, there are obvious risks to this as to any such abstraction. Kauffman's use of adaptive landscape models is an abstraction designed to explore one kind of limit to adaptation through selection. As Kauffman is aware (and points out at the beginning of this discussion), in terms of the landscape model, there are two distinct kinds of limitation to fitness maximization. First, as he discusses, climbing 'upwards' on the landscape will generally take you to a local optimum where you will be 'trapped', and unable to approach the global optimum. But second, there are reasons for doubting whether selection will drive a population upwards on the landscape anyway.

Kauffman mentions two reasons for the possible failure of selection to move a population upwards on an adaptive landscape. First, if a population is undergoing selection at more than one locus, limits on recombination may prevent it from flowing up the fitness landscape defined by a particular locus; and second, the fitness of a particular phenotype does not have a fixed value, but will generally be a function both of its own frequency in the population, and of the frequency of other phenotypes in the population. Kauffman's treatment quite explicitly abstracts from these difficulties, and as I have emphasized, there need be nothing at all wrong with that. However, in this particular case I am inclined to question the legitimacy of this abstraction, and wonder whether consideration of these difficulties does not seriously threaten to undermine the coherence of the models to which it gives rise.

For consider what it would mean to take these features, particularly the second, seriously in terms of adaptive landscape models. Density dependence, that is, variation in fitness as a function of the frequency of a phenotype, will imply that as the population climbs an adaptive slope, the gradient of the slope changes. But more important still are effects due to changes in other features of the population. Quite independent evolutionary changes in the population will show up as exogenous changes in the landscape topology. Note that I am not thinking of epistatic interactions between genes — though these will also present problems — but interactions between the phenotypic determinants of fitness. (I might also add here, parenthetically, that I do find Kauffman's conclusions about likely limitations to the extent of epistatic interactions fascinating and provocative.) As the neck of the paradigmatic giraffe is lengthening, for example, there may be decreasing selective pressure for the capacity to digest twigs, since it can reach more leaves. We are led to think of another organism that once briefly evolved a similarly long neck, Alice. The landscapes we are considering will resemble the moving landscape of Wonderland, except that the landscape will sometimes move faster than Alice, and sometimes slower; and it will be moving in many directions. (Recall, lest we think that at least the picture has the virtue of being easily grasped, that we are dealing with multi-dimensional hyperspaces.) Whether such a multiply and variously shifting terrain can really be called a landscape at all can seriously be questioned.

Let me make a more constructive suggestion about where I think the fundamental problems with the adaptive landscape picture arise. I mentioned earlier that one of the main sources of trouble with adaptationism was the complexity of the relation between genetic and phenotypic properties. Kauffman very cunningly develops his models in the context of peptide sequences. On the one hand these are directly correlated with DNA sequences, and steps on the peptide landscape are directly correlated with well understood mutation events at the genetic level. On the other hand it is conceivable that changes in peptide sequence could have very direct fitness effects. Kauffman suggests that affinity to a hormonal receptor on a cell surface might be one type of effect directly correlated with fitness. But without denying that there might be cases with this simple structure, the case is a very special one. Fitness is, in the first instance at least, a property of phenotypes. It is, roughly, the disposition to survive and reproduce, and pace Dawkins, it is a disposition possessed, in the relevant sense, by phenotypes not genes. In special cases in which an immediate gene product has simple and direct effects on fitness it may be harmless to attribute fitnesses to particular DNA sequences; but in general I would question the legitimacy of attributing fitnesses to genes at all. Now Kauffman does not, on the face of it, do any such thing. But as I read him, the models he is discussing are really genetic fitness models, and the talk of peptide spaces is ultimately a plausible phenotypic disguise. Or at least, if they are to have relevance beyond these very special cases, this is the way they must be understood. Beyond such special cases, I think that more sophisticated (post-bean-bag, one might say) understandings of development threaten to undermine the very intelligibility of genetic fitnesses; and thus to undermine, except in very special cases, the intelligibility of the kind of fitness models that Kauffman describes.

What I have been arguing for in the last few pages might be summed up as follows. While Kauffman's analysis of adaptive landscapes is obviously and explicitly post-Panglossian, there is a sense in which it may still be adaptationist in ways that we should reject. Though the analysis is obviously not Panglossian, I suggest that the problems that have been raised in critiques of Panglossianism threaten to undermine even such more cautious analyses of the selective process. It probably seems that my argument here constitutes a total and nihilistic rejection of the entire theory of natural selection. This is not my intention, and I hope not implicit in what I am saying. I shall return briefly to this question at the end of these remarks, but first I shall say something about the ontogenetic aspect of Kauffman's paper.

I have less to say about Kauffman's discussion of self-organization in genetic nets, mainly because, as I have already indicated, I find the ideas exciting and provocative. I would just like to point out what I take to be a mutualistic relation between these speculations, and my sceptical remarks on adaptationism. On the one hand, as I proposed at the beginning of these comments, the difficulties with adaptationism have provided both motivation and space for such theoretical developments. But in addition, I take my sceptical remarks about adaptive landscapes to be strongly supported by Kauffman's specific views about ontogeny. In particular, it seems plausible that the picture Kauffman develops of numerous higher level effects — state cycle attractors — emerging from quasi-Boolean interactions between genetic elements, suggests a strongly bimodal distribution of effects of point mutations. If small mutations preclude the operation of a particular cycle, they will presumably have dramatic, typically lethal, effects. If they do not, as will perhaps be more common, they may have no significant effects. Put differently, Kauffman's scheme suggests a general way of understanding the puzzling phenomenon recognized for at least the last thirty years, of developmental homeostasis, the apparent insensitivity of the phenotype to very substantial genetic variability. If we think of development as primarily involving 'emergent' higher level properties of genetic nets, it is easy to imagine that many, perhaps most, small bottom level changes might be irrelevant. Perhaps it also illuminates the possibility of macro-mutations (hopeful monsters) with novel, conceivably useful, attractor cycles. At any rate, and to get to the point, if this is right, adaptive 'landscapes' will resemble neither the rolling hills of Kentucky nor the rugged mountains of Colorado, but will look more like Conan Doyle's Lost World: a plateau with precipitous cliffs on every side. It seems far removed from that in which each point mutation can be associated with a well defined, usually small, change in fitness.

A different reason why I find this part of Kauffman's work particularly exciting is the parallel it provides (remarked by Kauffman) with some of the ideas on cognitive science discussed by Francisco Varela. There is surely a close parallel between the neurophysiologist's Grandmother cell, and the grandmother's gene for taking care of her grandchildren — at least in respect of their non-existence. But it has generally been much easier to point out difficulties with traditional reductive methodologies than to provide alternative approaches for explaining the emergence of extremely complex and orderly phenomena from a jumble of neurons or DNA molecules. Perhaps, in the approaches we are now considering, we can see glimmerings of more productive ways of addressing these problems.

I said I would return to the question of whether I am proposing a totally nihilistic outlook on Darwinian evolution. Evolution, it is now customary to assert, will occur whenever a set of entities exhibits heritable variations in fitness. The problem that I have kept returning to in these remarks could be put at its baldest as follows: no class of biological entities systematically and reliably possesses heritable variations in fitness. Organisms exhibit lots of variation in fitness, and genes are superb exponents of heredity. But organisms, at least sexually reproducing ones, do a pretty poor job of inheritance, and the fitness of genes is a constantly fluctuating function of the genetic and macroscopic environment. Thus selection conceived either phenotypically or genetically rapidly runs into serious problems. Phenotypic approaches run into major, perhaps fatal difficulties over the limits to the heritability of the properties to which they appeal, whereas genetic approaches have great difficulty justifying the assumption that the entities they discuss are really objects of selection.

Despite this my position is not nihilistic. There is no doubt, to begin with, that there is enough correlation between genotypic and phenotypic variation for some evolution by selection to occur. This is amply illustrated by artificial selection experiments and classic natural examples such as industrial melanism. How *much* of evolutionary change can be explained by standard population genetic models is what remains to be determined. As Kauffman's concluding speculations illustrate, moreover, there may be very different kinds of selective process occurring than those presupposed by current theory. Much remains to be discovered about the history of life, and I think it would be dogmatic to make any firm presuppositions about the role that natural selection will play in the various accounts we may come to offer. Thus scepticism about, roughly, 'The New Synthesis', should not be taken to exclude the possibility that selection of some kind will prove to have a central role in some later, more adequate, account of evolution.

I want to conclude with two very general remarks, the first of which follows directly from what I have just been saying. That evolution by natural selection is the major explanation of the existence and diversity of terrestrial life is a proposition that continues to define by far the best developed current research program for addressing this fundamental question. But we should be on our guard against its becoming an unquestionable dogma. There is perhaps some pressure to adopt such a dogmatic position from the popular perception that only faith in such a dogma protects us from regressive and obscurantist positions such as special Creationism. But of course, the reverse is the case. The standard creationist argument is that one system of faith (theirs) is as good as another (evolution). It is by resisting dogmatism of any kind, and hence the characterization of the theory of evolution as an object of faith, that we can decisively refute such an argument.

Finally, and perhaps relatedly, I would like to suggest that we be cautious in our assumptions about what a scientific investigation of the origins of life may be able to deliver. According to certain still very widespread views of science, we should not ultimately be satisfied with any explanation short of a demonstration of the unconditional necessity — perhaps given certain initial conditions — of what we are aiming to explain. A strong form of such a tendency can be seen in the crudest forms of Panglossianism: what evolved had to evolve, because it was the best possible. But the more we explore biological evolution, it seems, the more possible sources of pure historical contingency we seem to discover. It must certainly be the goal of evolutionary biology to explain how the evolutionary processes that have occurred were even *possible*. But beyond that, there may be nothing more to tell than an inconceivably complex historical narrative.

Let me conclude by wholeheartedly endorsing the final statement of Kauffman's paper: "We want a true theory of biological evolution embracing self-organization, selection, and historical accident." And perhaps, I would only add, much else besides.

*Stanford University*

* The details of the problems with optimality are discussed by Philip Kitcher, Elliott Sober, John Maynard Smith, Richard Lewontin, and others in J. Dupré, ed., *The Latest on the Best: Essays on Evolution and Optimality*, Bradford Books/MIT Press, 1987.

DANIEL R. BROOKS

# INCORPORATING ORIGINS INTO EVOLUTIONARY THEORY

## INTRODUCTION

Evolutionary theory is experiencing a period of close scrutiny. Some authors assert that evolutionary theory is essentially complete (e.g. Charlesworth *et al.*, 1982; Stebbins and Ayala, 1981; Buss, 1987), and others argue for a replacement theory that is independent of Darwinian principles. What is needed, however, are approaches that try to integrate traditional principles and research programs with new ideas that can address problems not addressed by current evolutionary theory. This is exemplified by the titles of recent texts and articles that emphasize the need to 'expand' (Gould, 1980), 'finish' (Eldredge, 1985), 'extend' (Wicken, 1987), or 'unify' (Brooks and Wiley, 1988; Brooks *et al.*, 1989) evolutionary biology. Attempts to find common ground among these various proposals have begun (e.g. articles in Weber *et al.*, 1988). To my mind, two concepts are common to all these proposals. The first is the feeling that there is more to evolution than changes in gene frequencies in local populations under different environmental conditions. The second is that evolutionary theory has been preoccupied with questions of maintenance rather than with questions of origin of diversity. In an effort to complement the presentation by Kauffman (this volume), I will discuss some conceptual connections involving (1) the origin of biological order, (2) the origin of particular traits, (3) the origin of the environment in which natural selection takes place, and (4) the origin of natural selection as an evolutionarily relevant process.

Theories of biological evolution that stem from the neo-Darwinian synthesis share a Newtonian, or time-symmetrical, perspective (Depew and Weber, 1988). Natural selection is a local phenomenon that implies no large-scale temporal or spatial structure. In the synthetic theory of evolution, the genetic structure of a population is expected to remain at equilibrium unless acted upon by an external environmental force ('natural selection' broadly construed), which shifts the population to a new equilibrium. New environmental forces, however, can shift the



*Francisco J. Varela and Jean-Pierre Dupuy (eds), Understanding Origins*, 191–212.

genetic structure back to its previous equilibrium. Hence, the processes of microevolution postulated by the synthetic theory are in themselves time reversible. For example, suppose one were to view a film showing moths in a forest, and both the moths and the trees appear predominantly light-coloured at first, then become predominantly dark, and then predominantly light again. The observer could tell that natural selection had occurred but could not tell whether the film had been shown in the correct temporal sequence. There are, however many biological processes, such as reproduction, development (ontogenesis), death, speciation (phylogenesis), and extinction, that are inherently irreversible phenomena. The question of how time-asymmetric processes are to be explained and integrated into general biological theory is central to the new evolutionary biology.

Time-asymmetric processes can be of very different sorts. The regularity of stages in the 'evolution' of different kinds of stars is an example of time-dependent regularity without historical connection. The ontogeny of BO class stars appears to have been the same in every case. The similarity is due only to similar conditions and causes, not to a shared history. BO stars do not share a common ancestor that displayed the mass, luminosity and spectral characteristics that characterize BO stars. Rather, BO stars are formed when particular initial conditions in the prestellar stages of star development are realized. Other conditions, such as insufficient mass of prestellar gas, would lead to a different class of star being formed. In contrast to stellar evolution, biological evolution is dependent not only on initial conditions, but on the interplay of many unique events specific to the particular history of the evolving biological system. For example, many of the similarities shared by species are the result of common ancestry and not of the realization of a repeatable series of events originating from independent, but identical, initial conditions. Thus, such similarities (homologies) are embedded in an inherited, i.e. historical, matrix. In general, unlike many time-dependent physical systems, biological systems retain effects of historical events that are transmitted to them from other, similar predecessors (i.e., ancestral systems). Their time-dependent behavior unfolding into the future is constrained by events that happened in the past, the effects of which they carry with them. In some ways, biological evolution has more in common with the systems studied in cosmology than in the narrower subject of astronomy.

## ORIGIN OF BIOLOGICAL ORDER

Three attributes tend to distinguish living systems from non-living systems: (1) 'phase separation' between the 'inside' and the 'outside', (2) replication and reproduction and (3) a high degree of autonomy of biological systems from their environments. In living systems the rate and form of (re)production, development, and evolution may be affected by environmental fluxes, but is also dependent on internal production 'rules' ('information' or 'instructions'). These production rules determine much of the biological forms upon which environmental selection operates.

### *Energy*

Lotka (e.g. 1913, 1925) was among the first to discuss the nature of biological systems in terms of energy flows and energy partitioning. Biological systems are systems that persist in space and time by transforming energy from one state to another in a manner that generates structures which allows the system to continue to persist. There are two classes of energetic transformations that can be recognized in biological systems. The first type of transformation results in a net loss of energy from the system, the loss being measured as heat. Maurer and Brooks (submitted) called these *heat-generating transformations*. The second class of transformation changes unusable energy into states that can be stored and used in subsequent transformations. Maurer and Brooks (submitted) called these *conservative transformations*. All conservative transformations in biological systems are coupled with heat-generating transformations. Lotka (1913) suggested that if conservative energetic transformations could accumulate enough bound energy, the system could delay the time of its inevitable loss of structure due to heat-generating transformations. Under this view, biological systems could be considered systems that slow the rate at which energy stored by conservative transformations is degraded by heat-generating transformations.

Energy flows within biological systems are coupled with the production of entropy. Heat generating transformations generate thermal entropy, which can be taken as the tendency of the system to move towards disorganization. Conservative transformations produce structural entropy, which is a measure of the tendency of the system to move towards complexity. Dissipative structures (Prigogine, 1967; 1980) are non-equilibrium thermodynamic systems in which macroscopic order (structural entropy) is produced by dissipative processes that allow a lower rate of entropy production than if the processes were completely thermal. Because energy stored by conservative transformations degrades at a rate slower than the heat liberated during heat producing transformations, there is a period of time during which the system accumulates structural entropy. This time lag allows the system to isolate processes occurring within the system from those occurring outside the system. Thus, the system can be viewed as maintaining a phase separation between the processes occuring within the system and those occuring without. That is, fluctuations in processes occurring outside the system that could lead to disorder if carried out within the system are isolated from processes occurring within the system. Entropy changes ($dS$) in such systems can be decomposed into two components, one measuring exchanges between the system and its surroundings ($d_eS$) (entropic changes in the environment) and the other measuring production by processes internal to the system ($d_iS$) (entropic changes in the system). These systems must produce entropy internally ($d_iS > 0$). Or,

$$dS = d_eS + d_iS, d_iS > 0$$

The vast majority of entropic behavior for biological systems involves the exchange component ($d_eS$), and is manifested by the physical degradation of matter and energy taken from the environment by living systems and returned to it. Hence, entropy increases for exchanges are manifested in the environment of the organism. This is accompanied by an accumulation of organized and diverse biomass within the biological system according to autonomous production rules. These production rules must result in entropic behavior ($d_iS > 0$), even though they pertain to a relatively small fraction of the total energetic behavior of biological systems. How can order within the system originate from positively entropic behavior?

A class of open thermodynamic systems called dissipative structures can arise kinetically when the internal dynamics of the system change the system faster than it can equilibrate with its surroundings. Dissipative structures can also arise physically, when the boundary conditions are such that there is a physical barrier between the system and its surroundings. In each case, a phase separation between the system and its surroundings is formed, allowing the possibility of internal production rules that are not governed directly by fluxes from the environment, but by entropy production within the system. Retained entropy production, such as heat, could establish conditions allowing reactions to occur internally that would be impossible otherwise. The greater the phase separation, or distinction between system and surroundings, the greater the autonomy of the internal production rules. Biological membranes are maintained kinetically and produce a physical phase separation between the living system and its environment. Hence, biological systems behave as dissipative structures. Classical natural selection occurs precisely because of the existence of internal autonomous production rules in biological systems.

Production rules in biological systems are those processes for which there is an energetic 'cost' or 'allocation', even if that allocation is a small fraction of the total energetics of the system. Following Prigogine and Wiame (1946) and Zotin and co-workers (e.g. Zotin and Zotina, 1978), Brooks and Wiley (1988) denoted such allocations using the symbol $\Psi$, signifying a specific dissipation function. $\Psi$ includes at least two classes of processes: (1) heat-generating reactions, called the *external dissipation function* ($\Psi_{\alpha}$) and (2) conservative reactions, called the *bound dissipation function* ($\Psi_{\mu}$). For biological systems, $\Psi_{\mu}$ can be decomposed into the allocations for accumulating biomass ($\Psi^{b}_{\mu}$) and allocations for accumulating genetic diversity ('instructional information' of Brooks and Wiley, 1988) ($\Psi^{i}_{\mu}$). Hence,

$$d_{i}S = \Psi_{\alpha} + \Psi^{b}_{\mu} + \Psi^{i}_{\mu}$$

Maurer and Brooks (submitted) suggested that if all three components of $\Psi$ are included in the thermodynamic production term $d_{i}S$, biological systems exhibiting order must have the following properties: (1) the rules (although not necessarily the details) for both heat-generating and conservative transformations must be encoded in the macromolecular structure of the system, (2) those production rules must include 'information' or 'instructions' leading to non-random exchanges between the system and its surroundings, and (3) production by the conservative processes must be positively entropic. Under this view, the thermodynamic drive in biological systems results from production, which includes processes that result in the accumulation according to intrinsic production rules results in the non-random mechanical and chemical gradients within biological systems. The flow of free energy and of entropy increases are directly, rather than inversely, proportional.

Frautschi (1988; see also Layzer, 1975; Frautschi, 1982; Landsberg, 1984a, b) recently contrasted two classes of processes that generate entropy. The first is equilibration of temperatures between system and surroundings; for open systems this comes through heat-generating transformations. Biological systems exhibit this kind of entropic behavior through external dissipation processes ($\Psi_a$). The second is expansion of the phase space of the system, an increase in its number of accessible microstates (possible configurations). System organization increases so long as equilibration (equiprobable distribution of system over all of its microstates) takes longer than phase space expansion, allowing a lag between the realized entropy and the maximum possible entropy (Figure 1). New microstates are formed by the production of

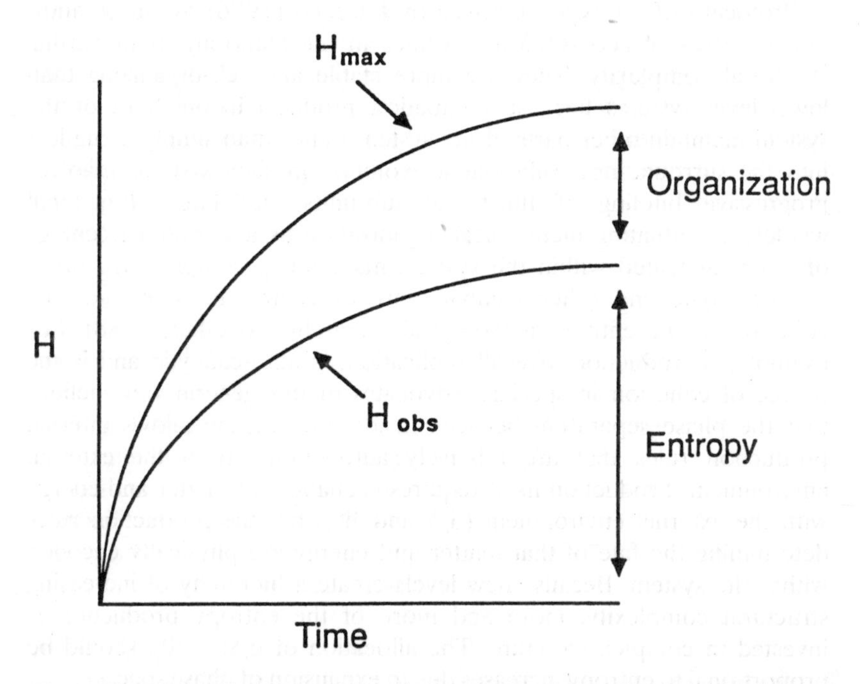


Fig. 1. The relationship between increasing entropy maximum ($H_{max}$) and the observed entropy ($H_{obs}$) of a physical system over time. The difference is proportional to organization while the value of $H_{obs}$ is a measure of the entropy of the system.

new components, either at a given organizational level or through the opening up of new levels. Free energy and entropy may accumulate together in such systems. Cosmological models of the expanding universe discussed by Frautschi (1982) explain the origin of free energy in the universe in this way. In biological systems this is accomplished by conservative transformations. For example, auto-catalytic processes producing monomers make 'monomer space' available for chemical evolution. Some monomers have high chemical affinities for each other, and will spontaneously clump into oligomers and polymers. A specific set of polymers can correspond to different distributions of monomers, depending on how the clumping occurred. Thus, some monomer microstates correspond to polymer macrostates. Once polymers begin to form, 'polymer space' becomes available to the evolving system. Causal interactions between polymers create new levels or organization, and so on.

Protocells of the type discussed by Lazcano (1986) would be additional classes of accessible microstates in the hierarchy of increasing structural complexity. They are more stable and self-organizing than lower-level systems because dissipation products in one part of the system maintain other parts of the system rather than simply being lost into the surroundings. Subsequent evolution in such systems involves progressive binding of functional sub-units into larger functional wholes, accentuating hierarchical organization as a greater percentage of $d_iS$ is allocated within the system, maintaining complex structures. Auto-catalytic and other feedback processes are the source of the cohesion of the entities in biological hierarchies (Collier, 1988). For example, reproduction, like all replication, is auto-catalytic and is the source of cohesion in species. Advocates of this general view believe that the phase separation between inside and outside allows internal production rules that are relatively autonomous from the external environment. Production itself requires exchanges of matter and energy with the external environment ($d_eS$ and $\Psi_a$) but the production rules determining the fate of that matter and energy are physically encoded within the system. Because new levels create a hierarchy of increasing structural complexity, more and more of the entropy production is invested in complex structure. The allocation of $d_iS$ to $\Psi_\mu$ should be proportional to entropy increases due to expansion of phase space.

*Information*

Reproduction and replication transmits semi-conserved copies of these production rules. Because the primary manifestation of these rules is ontogenies, Waddington (1967) referred to them as 'instructions', although they are more commonly thought of as 'information'. On this view, living systems could be called 'informed auto-catalytic systems' (Wicken, 1987; Brooks and Wiley, 1988). For this reason it has been useful to develop information theoretical analogs of the expanding phase space process (Brooks *et al.*, 1984; Brooks *et al.*, 1988; Brooks and Wiley, 1988; Smith, 1988).

Information depending on the state of a physical system is called *bound information* (Brillouin, 1962). It is information in the sense that the physical system has a determinate form. Measurements read this information, replicating it in a representation of the system. Genetic information is bound information, but also serves as a code for information transmission. Configurations with both these characteristics are called *arrays* (Collier, 1986; Brooks and Wiley, 1988). Arrays are composed of relatively stable elements that can combine freely or under moderate constraint (like the letters of the alphabet). A set of elements and the arrays they can form constitute a physical information system. Cohesion creates new higher level elements which can form their own arrays, whose microstates are the lower level arrays.

This distinction between informational macrostates and microstates yields a physically based informational entropy determined entirely by conditions within the information system itself. If causal process can be regarded as computations, algorithmic information theory (Kolmogorov, 1968; Chaitin, 1975) is the best way to show this. Algorithmic information theory yields the same information values as the more common probabilistic and combinatorial approaches, except for a small additive factor for computational overhead. The algorithmic information content $I(X)$ of an object $X$ is the size (in bits) of the shortest program that computes $X$. The entropy of a macrostate $M$, whether an array or other physical state, is the indeterminacy of its microstate $m$ given $M$. This is just the conditional information $I(m/M)$ of $m$ given $M$, which is defined as the length of the shortest program required to compute $m$ given $I(M)$. In this case $I(m/M) = I(m) - I(M)$. Because causal processes involving only properties of $M$ cannot compute states with information greater than $I(M)$, $m$ is random with respect to $M$ to the extent that it is underdetermined by $I(M)$, i.e. to degree $I(m/M)$, the entropy of $M$. The emergence of macrostates through processes producing cohesion, then, creates entropy unless only one microstate is possible for the new macrostate. (For further details, see Collier, 1986; Brooks *et al.*, 1988; Brooks and Wiley, 1988; Smith, 1988).

For a given level in a physical information hierarchy, the difference between the entropy maximum ($H_{max}$) and the actual entropy ($H_{obs}$) measures the organization of the system at that level (Figure 1). This difference is $I(M)$, the macroscopic information (Layzer, 1975) (Figure 2). It is also a measure of constraint, representing possible variation that has been historically excluded (Figure 3). The actual entropy ($H_{obs}$, Figures 1—3) is the internal entropy ($\Psi_{\mu}$, the bound dissipation) of the physical information system. It is also a measure of complexity (Figure 2), or realized diversity (Figure 4) of the system.

The dynamic presented heuristically in Figures 1—4 has three major elements: (1) $H_{obs}$ is an increasing function of time, as mandated by the

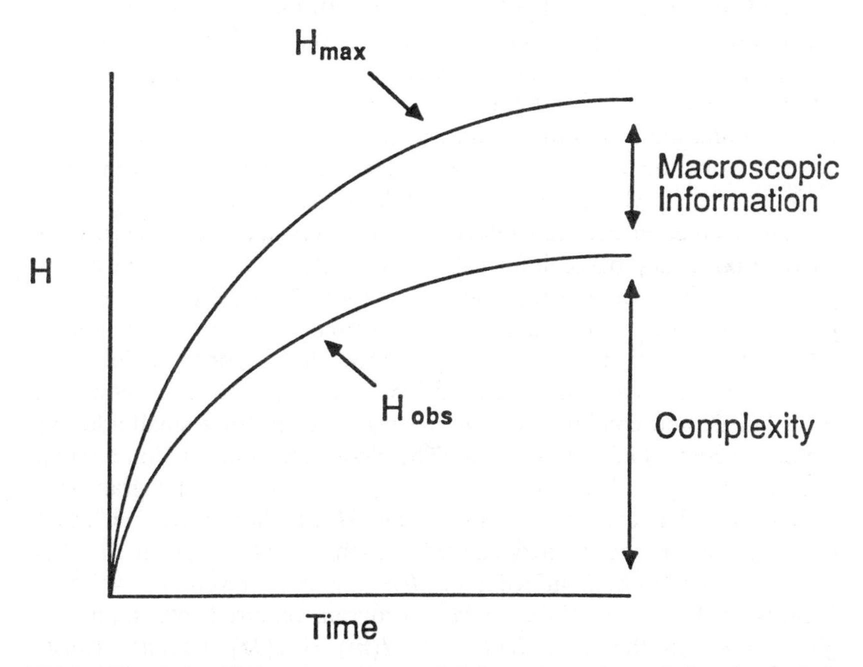


Fig. 2. The relationship between macroscopic information and complexity of a physical information system.

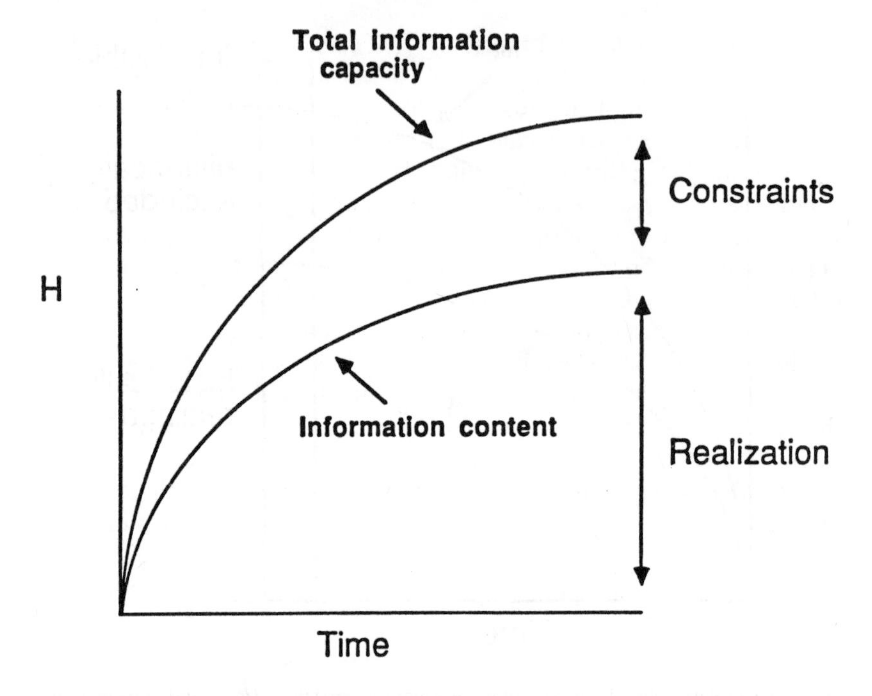


Fig. 3. The relationship between total information capacity ($H_{max}$) and information content ($H_{obs}$) of a physical information system. The difference between total information capacity and information content is proportional to the constraint placed on the information system. The entropy of information ($H_{obs}$) gives a measure of the amount of realized information diversity.

Second Law of Thermodynamics; (2) $H_{obs}$ is a concave function of time, as historical constraints retard the rate of entropy increases; and (3) the difference between $H_{max}$ and $H_{obs}$ is an increasing function of time, proportional to the growth of organization in the system. Smith (1988) has shown that a general class of mathematical models, called 'partitioned Lebesgue spaces with automorphism', has these properties. This class of models includes stationary Markov chains, which population biologists have long found useful in simulations of microevolutionary processes. Furthermore, the behavior of representatives of this class of systems can be documented using *Hierarchical Information Theory* (Brooks *et al.*, 1988) as modified by Smith (1988), as an accounting system.

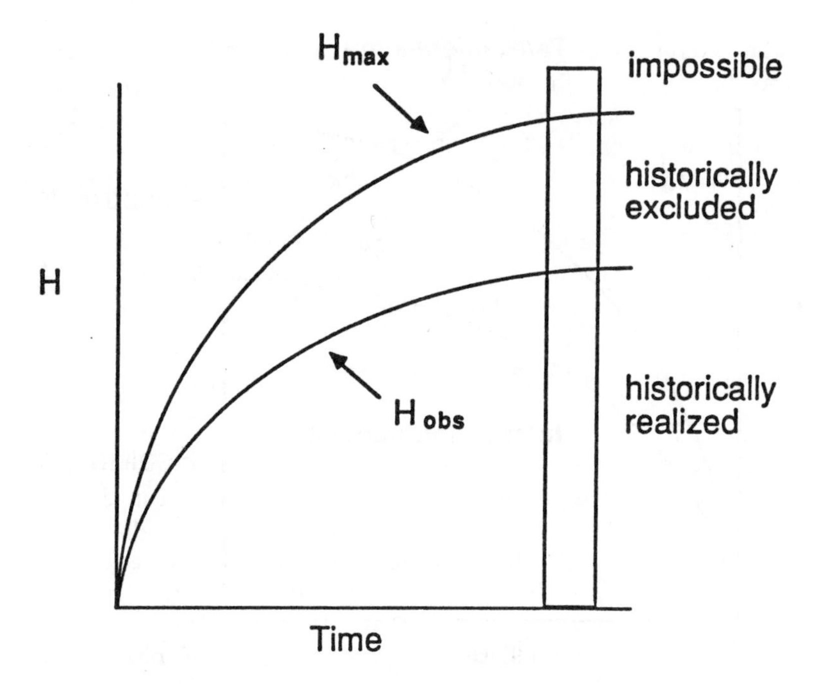


Fig. 4. The relationship between total information capacity ($H_{max}$) and information content ($H_{obs}$) of an array of physical information systems comprising a number of evolutionary lineages. Historically realized diversity is measured by $H_{obs}$. Historical exclusion of the expression of certain kinds of information is proportional to $H_{max} - H_{obs}$. The area above $H_{max}$ represents impossible combinations at a given time. Note that what is impossible at one time period may become possible at a later time period. Also note, however, that evolution is characterized by an ever-increasing area of historically excluded combinations.

## THE INTERACTION OF GENEALOGY AND ECOLOGY IN EVOLUTION

Environmental and genealogical phenomena are intimately connected in biology. Pre-biotic environmental conditions established the boundary (characterized by Salthe, 1985, as a pre-biotic ecology) within which life could originate. Conversely, genealogical processes that characterize life and evolution are autonomous enough from environmental conditions to be capable of overrunning available required resources and of changing the environment substantially. The longer life exists on this planet, the more it shapes the environment. Today, much of the environment consists of the products of genealogical processes. Eldredge and Salthe (1984), Salthe (1985), and Eldredge (1985, 1986) have suggested two forms of hierarchically-organized behavior in biology. The realm of *interactors* is the *ecological hierarchy*, and encompasses exchanges of matter and energy between system and environment ($d_eS$). The realm of *replicators* is the *genealogical hierarchy*, resulting from production processes ($d_iS$).

### *Orderly Diversification of Biological Form*

The dynamics of information in genealogy determines a 'phase space' that is filled to the extent that different possible genetic combinations are realized. It grows entropically in proportion to the accumulation of novelties arising through changes in genetic systems. Reproduction and recombination tend to fill that space. That the phase space is only partially filled tells us that there are constraints on entropy increases, either inherent constraints, such as those arising from the 'rules' of the genetic system (such as those discussed by Lima-de-Faria, 1983) or extrinsic forces. But this does not describe the particulars of the constraints. What is needed is an inventory of 'assembly rules' for different levels of organization in biological systems, analogous to the set of assembly rules provided by chemical kinetics for physico-chemical systems.

Goodwin (1985 and references therein) outlined an approach to theoretical biology called 'structuralism'. Goodwin proposed that there is a dichotomy between contingencies, which he associated with phylogeny, and regularities, constraints, or order in biological systems, which he associated with 'generative principles' derived from field theories of morphogenesis. The generative process that underlies all of biology is thought to determine discrete 'morphological stability domains' (Goodwin, 1985; p. 117) that will occur regardless of ancestry or environment. Evolution is "intelligible only in relation to the logic of the creative process" (Goodwin, 1985, p. 118). Structuralism of this form attempts to reduce biological explanations to the kinds of explanations usually associated with physics and chemistry. Such explanations tend to be deterministic and time-independent with respect to the expectations of natural laws. Hence, the structuralist approach views species as natural kinds rather than individuals. In a similar vein, Goodwin (1982) suggested that the only rational taxonomy would be a periodic table of morphogenetic fields rather than a phylogenetic hierarchy of species.

Waddington (1967, p. 109) considered morphogenetic fields to be "something that is extended in the time dimension." He viewed developmental phase space as an epigenetic landscape characterized by 'chreodes' or developmental trajectories. A chreod for Waddington consisted of a "region in a multi-dimensional configuration space. The region is extended along the time dimension and along the three axes of space, and also includes a number of other dimensions . . . . Such a region has a chreodic character when there is a hyperspace within it, extended in the time dimension, which acts as an 'attractor' with respect to the neighboring vector fields." I believe that many biological processes which are characterized by regularities, constraints and order are also inherently historical, i.e. time-dependent, in nature. Among these are ontogeny, reproduction, and speciation. Unlike Goodwin, I do not see a dichotomy between historical effects and regularities in biology — historical effects are part of the regularities. Nor do I see a conflict between historical explanations and physical law, since in physics and chemistry time-dependent phenomena are explained by reference to entropic phenomena.

There is a second view of the relationship between ontogeny and phylogeny that is relevant to this discussion. To some, ontogeny and phylogeny are parts of the time-dependent nature of biology, but differ markedly as processes. Ontogeny is viewed as predictable and irreversible, a dynamical macroscopic process that smooths out perturbations resulting in a high degree of predictability in the outcome. Phylogeny, to the contrary, is unpredictable and involves only the relatively passive accumulation of genetic diversity resulting from the interplay of ontogenies and environmental changes (Salthe, 1985; Wicken, 1987). On this view, phylogeny could be viewed as an epiphenomenon having no dynamical behavior or regularities. I believe that the concept of temporal scaling provides a common ground for reconciling these apparent differences in perspective. In this view, ontogeny and phylogeny are both products of the genealogical hierarchy and have similar dynamics (see empirical examples in Brooks and Wiley, 1988), differing primarily in their temporal and spatial scaling. Ontogeny involves the replication and modification of cells and operates at the level of individual organisms (unicellular or multi-cellular) over relatively short time scales. Phylogeny involves the replication and modification of genotypes and operates at the level of the individual species (unicellular or multicellular) over long time scales. Both processes involve (1) irreversibility, (2) a mixture of historical constraints that provide predictability and indeterminism that provides individuation, and (3) spontaneous growth, although the tokens of growth are different. Both processes are dynamical and their units are capable of being acted upon by environmental selection.

The bulk of constraints on entropy increases in evolution originate in the genealogical flow of information, transmitted through reproduction. History acts as a 'force' (i.e. constraint) keeping biological systems from occupying all possible genetic configurations. Two major manifestations of biological production rules that evolved long ago are membranes and organelles. Membranes increase the phase separation between system and surroundings (as discussed earlier), sequestering energy and allowing new production rules. The more membranes, the more times energy can be used before it is lost from the system (i.e. the greater the entropy production). Organelles (whether derived endosymbiotically or via membrane elaboration) provide new production rules, but organelles are allowed only if enough energy to maintain their structure can be sequestered (e.g. by the production of membranes). Membranes restrict the influx of matter and energy, but need both to be maintained. There is probably an 'upper bound' on structural complexity, given that the structure must permit enough influx to 'pay for itself' energetically. Other evolutionary trade-offs are also implied: (1) there will be increasing autonomy of internal production rules, leading to increasing amounts of endogenous organization, but this will be offset by a restriction of options and loss of flexibility in the face of environmental changes; (2) the endogenous organization will provide historical continuity and inherent constraints on evolutionary change, but change will occur because production rules are entropic; (3) increasing variation will increase the kinds of energy-using systems in the biosphere, but not all will survive because of the organization of the biosophere, leading to an expectation that environmental selection will be an important component of evolution; and (4) the elaboration of membranes and membrane properties will allow the aggregation of subsystems (e.g. uni-cells) into hierarchically organized systems (i.e. multicellular organisms), but this will be offset by a loss of evolutionary independence for the component subsystems.

One term for this approach to evolutionary explanations is *historical structuralism* (Brooks and O'Grady, 1986; Brooks and Wiley, 1988). By documenting the historical origins of traits, either functional or structural, workers in this new research program hope to find, among other things, the origins of 'key innovations' that arose once historically and served as major constraints on subsequent evolutionary diversification, and to find recurring themes in biological evolution implicating structuralist regularities in the evolutionary process (Lauder, 1981, 1982). Phylogenetic systematics (Hennig, 1966; Wiley, 1981), vicariance biogeography (Nelson and Platnick, 1981; Wiley, 1988a, b), historical ecology (Brooks, 1985; Wanntorp *et al.*, 1990; Brooks and McLennan, 1991), and the phylogenetic study of heterochrony (Fink, 1982) are research programs in historical structuralism. A recent case study (Wake and Larsen, 1987) emphasizes the connections between this approach to evolutionary explanations and more traditional approaches.

### *Origin of the Ecological Hierarchy*

Ecosystems have long been considered energy-exchange systems (Lotka, 1925; Lindeman, 1942, Emig, 1985; Maurer, 1987). Lindeman (1942) suggested that production in ecosystems involved both 'dissipation' (loss) and 'growth' (biomass accumulation). Over very short time periods ecosystems behave like dissipative structures and can be treated essentially as equilibrium phenomena (see e.g. Levins, 1975). He suggested that over long time periods, the macroscopic state of an ecosystem was proportional to the ratio of dissipation: growth. A value of unity corresponds to a steady-state ecosystem, with some (theoretical) probability of extinctions for particular members. Values higher than unity indicated senescent ecosystems, with an increased probability of extinctions; values lower than unity indicated a growing ecosystem, with decreased probabilities of extinction. Recently, Ulanowicz (1980, 1986) has attempted to provide a coherent conceptual framework for understanding this aspect of ecosystem behavior.

The unified theory suggests that some aspects of eco-systems are highly constrained genealogically. Thus, while the exchange of energy and matter with the environment is essential to biological functioning, the irreversible behavior and the increasing complexity characteristic of biological evolution is not an unaided consequence of environmental forces. The creation and maintenance of biological systems requires environmental resources but does not require that the information in those systems originates in the environment (although sometimes it may). Biological systems have intrinsic capacities to create hierarchically organized structures if they have an adequate source of matter and energy. The environment is not inherently organized as the ecological hierarchy; its organization into an ecological hierarchy is largely a consequence of organization intrinsic to the genealogical hierarchy. Or, in other words, species do not fill empty niches, they create their own niches. At the same time, the environment provides an important constraining influence on biology, and the (self-generated) ecological hierarchy plays an important feedback role in evolution (see next section). The ecological hierarchy is the means by which two different genealogies, or two different generations in one genealogy, can causally influence one another. The 'adaptive landscape' fits organisms so well because it is created by the organisms themselves.

If the entropic behavior of a system affects its surroundings, then it is possible for the system to play a role in structuring its environment. For biological systems,

$$\mathrm{d_i}S = \Psi = \Psi_{\alpha} + \Psi^{\mathrm{b}}_{\mu} + \Psi^{\mathrm{i}}_{\mu}$$

This production term corresponds to the genealogical hierarchy. The products of genealogical processes exist by exploiting 'entropy gradients' in the surroundings, which we view as the ecological hierarchy, and associate with $\mathrm{d_e}S$. These gradients, and thus the ecological hierarchy, are determined partly by abiotic factors and partly by biotic factors. $\Psi_{\alpha}$ determines how the abiotic portion of the ecological hierarchy can be structured by products of the genealogical hierarchy. In this way part of the production term, $\mathrm{d_i}S$, can influence the exchange term, $\mathrm{d_e}S$. $\Psi^{\mathrm{b}}_{\mu}$ and $\Psi^{\mathrm{i}}_{\mu}$ determine the portions of the ecological hierarchy that are products of the genealogical hierarchy. The relationship between the two hierarchies based on the above view is shown schematically in Figure 5.

## ORIGIN OF NATURAL SELECTION

Over long ('evolutionary') time intervals macroscopic information ($I$), or organization, should accumulate in biological systems (Figure 1). Therefore, values of the function $Q$, where

$$Q = 1 - (H_{\mathrm{obs}}/H_{\mathrm{max}}) = I/H_{\mathrm{max}}$$

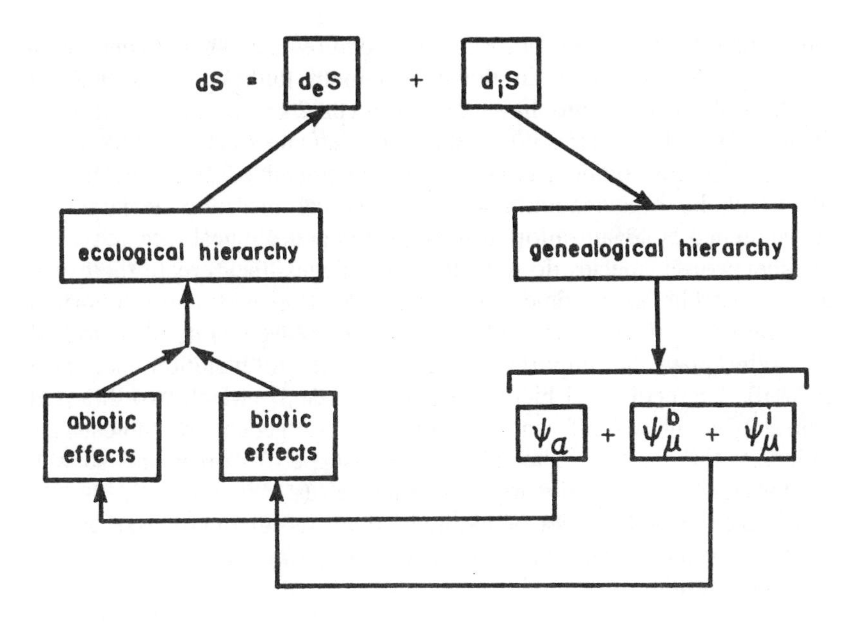


Fig. 5. The conceptual relationship between the genealogical hierarchy (biological production) and the ecological hierarchy (environment/biological systems exchanges) in terms of entropic behavior of open systems. The degree to which the genealogical processes shape the ecological hierarchy is the extent to which organisms have changed the environment of earth during evolution.

should increase over time. Landsberg (1984b) termed *Q macroscopic order*. It is also known as *redundancy* (Gatlin, 1972). The redundancy concept also figures in communications theory views of information (Shannon and Weaver, 1949). In that formulation, however, the definition of redundancy relies on the interpretation of a message by the source. If we use the assumptions of Collier (1986) about physical information systems, then the redundancy concepts are the same. To do so, we must recognize that the message, the source of transmission, and the receiver of the message are parts of the same system. We believe that this underlies at least some attempts to use information theory to help with studies of communication in animal behavior (Hailman, 1977). If this dynamic in biological systems results from intrinsic (genealogical) rather than extrinsic (environmental) factors, we should expect that the products of reproduction and ontogeny to show high degrees of structural and functional redundancy without being literal representations of the environment. If the environment is the source of macroscopic organization, the mutual information of biological systems $B$ and their environment $E$, denoted as $I(B:E) = I(B) - I(B/E) = I(E) - I(E/B)$, should be a relatively high proportion of $I(B)$. We assume that $I(B:E)/I(B)$ is low. If this is the case, it is possible for reproduction to produce more organisms needing a particular environmental resource than is available in the environment. When this happens, some organisms that are otherwise functional will not survive to reproduce or will not reproduce to the same extent as others. If environmental conditions cause differential reproduction of different genotypes, we speak of natural selection in its classical environmental mode. Thus, natural selection is a mechanism of environmental effects under conditions established by genealogically driven self-organization. Selection does not introduce genealogical organization, it reduces excess genealogical organization. In doing so, it may reinforce or add to ecological organization.

Selection-induced adaptation (ecological organization) reduces genealogical diversity, reducing the part of $I(B)$ in $I(B/E)$, increasing $I(B:E)/I(B) = 1 - I(B/E)/I(B)$. $I(B:E)/I(B)$ is maximal when $I(B:E) = 0$, i.e., when $B$ is determined entirely by $E$. The limit of adaptation produced solely by selection, then, is loss of phase separation between organism and environment, i.e. death or extinction. This extreme consequence suggests that production through increases in $I(B:E)/I(E) = 1 - I(E/B)/I(E)$ must balance or dominate losses through increases in $I(B:E)/I(B)$ in a generally fit population. If we consider the whole adaptive process of introducing and reducing organization, increases in $I(B:E)/I(E)$ are facilitated by entropy producing increases in the genetic phase space which permit increases in $I(B:E)$. The overall entropy increase is constrained by limitations on resources, and it may be directed by selection as well. Thus, adaptation is constrained by genealogical as well as environmental factors.

Genealogical processes determine constraints on the manner in which and the degree to which populations respond to environmental influences. I believe therefore that more evolutionary information for any group of organisms is carried in their genealogies than in their environments. This can be shown by a simple thought experiment. Suppose we were to pick, at random, any organism from a known tide pool. Then we would pick, again at random, a crab from an unknown place somewhere in the world. If we then asked for a list of predicted morphological, behavioral, and ecological characteristics of the unknown organism from the known habitat (the tide pool) and of the unknown crab from an unknown habitat, we would expect more of the predictions for unknown crab to be correct. Knowing that we are dealing with a crab imparts more evolutionary and ecological information than the most detailed description of the tide pool. In terms of research programs in evolutionary biology, this means that all evolutionary explanations, including those for ecological traits and interactions, must include explicit reference to genealogical sequences (phylogenies) that must extent back to the origin of the relevant traits (see e.g. Lauder, 1981, 1982).

## CONCLUSIONS

Proposals for an expanded agenda is based on trying to find a common theme to link various research efforts, at least conceptually. The theme that I have emphasized herein is the use of energy in maintaining and transforming ordered states of matter. Using information and entropy concepts associated with discussions of energy use, as an accounting system, a common phenomenology for a number of developmental and evolutionary processes occurring on different levels of organization in biological systems has been documented (Brooks and Wiley, 1988). Processes relevant to evolution are either generative (originating or diversifying) in their effects, or conservative (maintaining or selective) in their effects. The interplay of diversity-promoting and diversity-limiting processes through time produces historically-constrained order. Many processes affect biological systems, at all levels of organization and at all times, but their effects are often manifested on different time scales. Changes occurring on times scales shorter than speciation rates will appear as microevolutionary patterns; those occurring on time scales longer than speciation rates will appear as macroevolutionary patterns. In this sense, macroevolutionary processes are neither redicible to, nor autonomous from, microevolutionary processes. The unified theory of evolution is not a non-Darwinian theory, although it can be construed as extending neo-Darwinian theory to the extent that neo-Darwinian theory is preoccupied with reducing biological explanation to changes in gene frequencies in populations subjected to different environmental selection regimes.

## ACKNOWLEDGEMENTS

I would like to express my appreciation to Dr. John Collier, Department of Philosophy, University of Calgary, Dr. Brian Maurer, Department of Zoology, Brigham Young University, Dr. Jonathan Smith, Department of Mathematics, Iowa State University, and Dr. E. O. Wiley, Museum of Natural History, University of Kansas for fruitful discussions. This work was supported by operating grant A7696 from the Natural Sciences and Engineering Research Council of Canada.

*Department of Zoology,*
*University of Toronto*

## REFERENCES

Brillouin, L.: 1962, *Science and Information Theory*. New York: Academic Press. 2nd ed.

Brooks, D. R.: 1985, 'Historical ecology: A new approach to studying the evolution of ecological association', *Ann. Missouri Bot. Garden*, **72**, 660—680.

Brooks, D. R., Collier, J., Maurer, B. A., Smith, J. D. H. and Wiley, E. O.: 1989, 'Entropy and information in evolving biological systems', *Biol. Philos.*, **4**, 407—432.

Brooks, D. R., Cumming, D. D. and LeBlond, P. H.: 1988, 'Dollo's Law and the Second Law of Thermodynamics: Analogy or extension?' in: *Entropy. Information and Evolution: New Perspectives on Physical and Biological Evolution*, Weber, B., Depew, D. J. and Smith J. D. (Eds.). Cambridge: MIT Press, pp. 189—224.

Brooks, D. R., LeBlond, P. H. and Cumming, D. D.: 1984, 'Information and entropy in a simple evolution model', *J. Theor. Biol.*, **109**, 77—93.

Brooks, D. R. and McLennan, D. A.: 1991, *Phylogeny, Ecology and Behavior: A Research Program in Comparative Biology*. Chicago: Univ. Chicago Press.

Brooks, D. R. and O'Grady, R. T.: 1986, 'Non-equilibrium, thermodynamics and different axioms of evolution', *Acta Biotheror.*, **35**, 77—106.

Brooks, D. R. and Wiley, E. O.: 1988, *Evolution as Entropy: Toward a Unified Theory of Biology*. Chicago: Univ. Chicago Press. 2nd ed.

Buss, L.: 1987, *The Evolution of Individuality*. Princeton: Princeton Univ. Press.

Chaitin, G. J.: 1975, 'A theory of program size formally equivalent to information theory', *J. ACM*, **22**, 329—340.

Charlesworth, B., Lande, R. and Slatkin, M.: 1982, 'A neo-Darwinian commentary on macroevolution', *Evolution*, **36**, 474—498.

Collier, J.: 1986, 'Entropy in evolution', *Biol. Philos.*, **1**, 5—24.

Collier, J.: 1988, 'Supervenience and reduction in biological hierarchies', *Can. J. Philos.*, suppl. volume., **14**, 209—234.

Depew, D. J. and Weber, B.: 1988, 'Consequences of nonequilibrium thermodynamics for the Darwinian tradition', in: *Entropy, Information and Evolution: New Perspec-*

*tives on Physical and Biological Evolution.* Weber, B., Depew, D. J. and Smith, J. D. (Eds.). Cambridge: MIT Press, pp. 317—354.

Eldredge, N.: 1985, *Unfinished Synthesis.* New York: Columbia Univ. Press.

Eldredge, N.: 1986, 'Information, economics and evolution', *Ann. Rev. Ecol. Syst.*, **17**, 351—369.

Eldredge, N. and Salthe, S. N.: 1984, 'Hierarchy and evolution', in: *Oxford Surveys in Evolutionary Biology*, Dawkins, R. and Ridley, M. (Eds.), Vol. 1, pp. 182—206.

Emig, C. C.: 1985, 'Relations entre l'espace, structure dissapatrice biologique, et l'écosysteme, structure dissapatrice écologique. *C. R. Acad. Sci. Paris*, **300**, 323—326.

Fink, W. L.: 1982, 'The conceptual relationship between ontogeny and phylogeny', *Paleobiology*, **8**, 254—264.

Frautschi, S.: 1982, 'Entropy in an expanding universe', *Science*, **217**, 593—599.

Frautschi, S.: 1988, 'Entropy in an expanding universe', in: *Entropy, Information and Evolution: New Perspective on Physical and Biological Evolution*, Weber, B., Depew, D. J. and Smith, J. D. (Eds.). Cambridge: MIT Press, pp. 11—22.

Gatlin, L. L.: 1972, *Information Theory and the Living System.* New York: Columbia Univ. Press.

Goodwin, B. C.: 1982, ' Development and evolution', *J. Theor. Biol.*, **97**, 43—55.

Goodwin, B. C.: 1985, 'Changing from an evolutionary to a generative paradigm in biology, in: *Evolutionary Theory: Paths into the Future*, Pollard J. W. (Ed.). London: John Wiley & Sons, pp. 99—120.

Gould, S. J.: 1980, 'Is a new and general theory of evolution emerging?', *Paleobiology* **6**, 119—120.

Hailman, J. P.: 1977, *Optical Signals: Animal Communication and Light.* Bloomington: Univ. Indiana Press.

Hennig, W.: 1966, *Phylogenetic Systematics.* Urbana: Univ. Illinois Press.

Kolmogorov, A. N.: 1968, 'Logical basis for information theory and probability theory', *IEEE Transactions on Information Theory*, **14**, 662—664.

Landsberg, P. T.: 1984a, 'Is equilibrium always an entropy maximum?', *J. Stat. Physics*, **35**, 159—169.

Landsberg, P. T.: 1984b, 'Can entropy and "order" increase together?', *Physics Letters*, **102A**, 171—173.

Lauder, G. V.: 1981, 'Form and function: Structural analysis in evolutionary biology', *Paleobiology*, **7**, 430-442.

Lauder, G. V.: 1982, 'Historical biology and the problem of design', *J. Theor. Biol.*, **97**, 57—68.

Layzer, D.: 1975, 'The arrow of time', *Sci. Amer.*, **233**, 56—69.

Lazcano, A.: 1986, 'Prebiotic evolution and the origin of cells', *Treb. Soc. Cat. Biol.*, **39**, 73—103.

Levins, R.: 1975, 'Evolution of communities near equilibrium', in: *Ecology and Evolution of Communities*, Cody, M. L. and Diamond, J. M. (Eds.). Cambridge, Massachusetts: Belknap Press, pp. 16—50.

Lima-de-Faria, A.: 1983, *Molecular Order and Organization of the Chromosome.* Amsterdam: Elsevier.

Lindeman, R. L. : 1942, 'The trophic-dynamic aspect of ecology', *Ecology*, **23**, 399—418.

Lotka, A. J.: 1913, 'Evolution from the standpoint of physics, the principle of the persistence of stable forms', *Sci. Amer.* suppl. **75**, 345—346, 354, 379.

Lotka, A. J.: 1925, *Elements of Physical Biology.* Baltimore: Williams and Wilkins.

Maurer, B. A.: 1987, 'Scaling of biological community structure: A systems approach to community complexity', *J. Theor. Biol.* **127**, 97—110.

Maurer, B. A. and Brooks, D. R.: submitted, 'Energy flow and entropy production in biological system', *J. Ideas.*

Nelson, G. and Platnick, N.: 1981, *Systematics and Biogeography: Cladistics and Vicariance.* New York: Columbia Univ. Press.

Prigogine, I.: 1967, *Thermodynamics of Irreversible Processes.* New York: Wiley-Intersci. 3rd ed.

Prigogine, I. : 1980, *From Being to Becoming.* San Francisco: W. H. Freeman and Co.

Prigogine, I. and Wiame, J. W.: 1946, 'Biologie et thermodynamique des phénomenes irréversibles', *Experientia,* **2**, 451—453.

Salthe, S. N.: 1985, *Evolving Hierarchical Systems: Their Structure and Representation.* New York: Columbia Univ. Press.

Shannon, C. E. and Weaver, W.: 1949, *The Mathematical Theory of Communicaton.* Urbana: Univ. Illinois Press.

Smith, J. D. H.: 1988, 'A class of mathematical models for evolutuion and hierarchical information theory', *Inst. Math. Appl. Preprint Series* **396**, 1—13.

Stebbins, G. L. and Ayala, F. C.: 1981, 'Is a new evolutionary synthesis necessary?', *Science,* **213**, 967—971.

Ulanowicz, R. E.: 1980, 'An hypothesis on the development of natural communities', *J. Theor. Biol.,* **85**, 223—245.

Ulanowicz, R. E.: 1986, *Growth and Development of Ecosystems and Communities.* New York: Springer-Verlag.

Waddington, C. H.: 1966, 'Fields and gradients', in: *Major Problems in Developmental Biology,* Locke, M. (Ed.). London: Academic Press, pp. 105—124.

Wake, D. B. and Larson, A.: 1987, 'Multidimensional analysis of an evolving lineage', *Science,* **238**, 42—48.

Wanntorp, H-E., Stearns, S. C., Brooks, D. R., Nilsson, T., Nylin, S., Ronqvist, F. and Weddell, N.: 'Phylogenetic approaches in ecology', *Oikos,* **57**, 119—132.

Weber, B., Depew, D. J. and Smith J. D. (Eds.).: 1988, *Entropy, Information and Evolution: New Perspectives on Physical and Biological Evolution.* Cambridge: MIT Press.

Wicken, J. S. 1987, *Evolution, Thermodynamics and Information: Extending the Darwinian Paradigm.* New York: Oxford Univ. Press.

Wiley, E. O.: 1981, *Phylogenetics: The Theory and Practice of Phylogenetic Systematics.* New York: Wiley-Interscience.

Wiley, E. O.: 1988a, 'Vicariance biogeography', *Ann. Rev. Ecol. Syst.,* **19**, 513—542.

Wiley, E. O.: 1988b, 'Parsimony analysis and vicariance biogeography', *Syst. Zool.,* **37**, 271—290.

Zotin, A. I. and Zotina, R. S.: 1978, 'Experimental basis for qualitative phenomenological theory of development', in: *Thermodynamics of Biological Processes.* Lamprecht, I. and Zotin, A. I. (Eds.). Berlin: deGruyter, pp. 61—84.

BRIAN C. GOODWIN

# THE EVOLUTION OF GENERIC FORMS

## INTRODUCTION

It is now widely accepted that if there are generative laws that operate during the development of organisms, then their morphologies will be constrained so that certain forms are possible and other are not. A possible consequence of this is that the hierarchical taxonomies of organisms arise not from dichotomous branching due to the historical winnowing process of natural selection, producing a discrete spectrum from an initial continuum, but from the intrinsic discontinuities that separate natural kinds generated by dynamical laws. If this is the case, then biological taxonomy has a basis not in the contingencies of history, but in the rational dynamics of biological organization. Such an eventuality would mean that Linnaeus was in a sense closer to the truth than Darwin, whose views on taxonomy were clearly expressed in such statements as "Our classifications will come to be, as far as they can be so made, genealogies; and will then truly give what may be called the plan of creation." It is now possible to construct such genealogies on the basis of, for example, the similarities and differences of genomic DNA sequences in different species. But such genealogies do not reveal the plan of creation. They do not even reveal the morphological relationships of organisms, because of the radical disjunction between the DNA content and structure of different species on the one hand, and their forms on the other. Evidently an understanding of the morphological relationships between species depends upon some other level of order than their DNA and their genealogical (historical) relationships, although is it perfectly clear that the former has some role to play in relation to form. However, the challenge to those taking the view that evolution at the morphological level reveals a temporal sequence of generic forms, intrinsically linked to the rational dynamics of development, is to describe both the nature of this developmental dynamics and why it should generate a hierarchical or Linnaean taxonomy. In this paper I shall explore certain approaches to this challenge.



*Francisco J. Varela and Jean-Pierre Dupuy (eds), Understanding Origins, 213–226.*

## METABOLIC AND DEVELOPMENTAL PATHWAYS

Matabolism and development have often been compared, notably by Goldschmidt and Waddington, two eminent explorers of the relations between development and evolution. I want to use the comparison to make a specific point about the relationships between general laws and their particular expression in organisms. What makes metabolic pathways *possible* is the differences in chemical potential between substrates and products. These thermodynamic relationships are measured under conditions of constant temperature and pressure, primarily by free energy differences. There is no way in which a metabolic sequence can run up a free energy gradient. Like the proverbial river that provides us with such a rich source of metaphors, metabolism always run downhill, products having less free energy than substrates. A metabolic sequence is a series of such downward steps, from one or more (relatively) stable metabolite(s) to the next. The set of possible metabolic sequences is determined by these thermodynamic properties.

However, the rate at which different steps in a possible sequence occur is dependent not upon the free energy difference between substrates and products, but on the activation energy involved in converting one metabolite into another. In the real world of process, rates are where the action is, and organisms control them by enzymes which reduce activation energies, and by ligands that secondarily affect an enzyme's ability to influence these energies. All of this is very familiar, and I apologize for the repetition of basic biological knowledge. However, what should be equally basic knowledge about morphogenesis is not so widely disseminated. In the metabolic case, there is a clear distinction between the laws of thermodynamics that make metabolic pathways possible, and gene products (enzymes) that alter rates of metabolic transformation. Consequently, gene products do not make metabolism possible; they stabilize particular expressions of the laws of thermodynamics in particular organisms by influencing specific transitions and by cross-linking rates in different pathways via control signals. The universal features of biochemistry are dependent upon basic chemistry and thermodynamics; the particulars arise from gene product specificities.

Now let us apply this argument to developmental pathways. Since I am concerned in this paper with morphogenesis, it is the shape-determining aspects of development that I want to consider. I shall argue that, just as gene products do not make metabolism possible, this being a result of physical and chemical laws, so gene products do not make morphogenesis possible, this being also a result of the laws of physics and chemistry. But what laws, exactly? Clearly, if we knew this in detail, then we would have a knowledge of morphogenesis as exact and rigorous as that of metabolism. But we already know enough to give a general answer. Since morphogenesis is about making shapes, the laws on which it is based are those that describe how forms are initiated in systems with particular types of space-time organization. Technically, these are the symmetry-breaking processes that result in the emergence of more complex from simpler (more symmetric) structures. In addition, it is necessary to have a general description of the basic building blocks out of which spatial forms are constituted, the analogues of metabolites which are the units of metabolic pathways. And finally, a complete description of morphogenesis requires some understanding of the energetic relations between these different building blocks, the spatial elements of morphogenetic sequences.

The first elements of a morphogenetic field theory are well-characterized. Spontaneous symmetry-breaking or bifurcation is the process in which a spatially uniform state, subject to random perturbation, develops into a stable non-uniform pattern as a result of the balance of forces acting within the system, which make the initial spatially-homogeneous state unstable. The first demonstration of how this could occur in a biochemical system was given by A. M. Turing (1952) in his celebrated paper 'The Chemical Basis of Morphogenesis'. In this he showed how enzyme-catalyzed biochemical reactions, together with diffusion, could result in an instability of spatially uniform states, which spontaneously transform into spatial patterns described by stationary waves of chemical concentration. This remarkable result showed how the laws of physics and chemistry, operating within a biological context, could generate patterns. The forces involved are those of chemical reaction together with diffusion. In relation to this theory, gene products (e.g. enzymes) do not themselves generate the patterns. They determine parametric values in the equations describing the potentially-bifurcating system and so can determine whether or not bifurcations occur, and they influence the wavelengths and amplitudes of any spatial patterns that arise.

This brings us to the second component of a theory of morphogenetic pathways: the elements out of which a morphogenetic sequence is constructed. Like all theories dealing with spatio-temporal organization, Turing's is a field theory, his equations describing the behavior of a reaction-diffusion field. There are many different types of field in physics and chemistry, each characterised by different equations and describing spatio-temporal patterns with distinctive features of waveform and rate of pattern initiation or transformation. But all field theories have certain properties in common: the solutions of such equations in their linearized forms are known as harmonic functions. These differ in wavelength, and any pattern of the field is initially described by some set of such harmonic functions. However, as pattern develops, the non-linear features of any particular field are expressed and distinctive wave-shapes emerge. In general this results in a discrete set of stable forms, solutions of the full field equations, which characterize the set of possible spatial patterns which a particular system can display. In the case of Turing's equations, there are many solutions, and a variety of examples has been described by Murray (1977, 1989) and Meinhardt (1982) in their analysis of biological pattern formation based on Turing's theory. These are the elements out of which any morphogenetic sequence would be constructed, if it is based upon this particular theory.

Finally, what about the energetic relations between successive steps in a pattern-forming process? This is the least well-characterized aspect of field theories, except in their simplest (linear) form, and in fact it is where the analogy with metabolic pathways begins to break down if taken too literally. I shall now give a morphogenetic example that illustrates the whole set of general principles discussed so far.

### *Axis Formation and Cleavage Patterns in the Amphibian Embryo*

The spherical amphibian egg when laid has a single axis, the animal-vegetal. It is not clear how this arises during oogenesis, but it appears to be a case of spontaneous symmetry breaking, the spherical oocyte generating an axis as a result of its own dynamics by a process at least qualitatively like that described by Turing. Fertilization of the egg by a sperm normally initiates the second symmetry breaking event, which generates a bilaterally symmetric embryo as revealed by the formation of the grey crescent on the side of the egg opposite to that of sperm entry. This grey crescent is the site where gastrulation will begin in the late blastula. That sperm entry is simply a nucleating event, a small stimulus that initiates a global response from a system in unstable equilibrium, is evident from the observation that other stimuli can override this and re-set the axis of bilateral symmetry. An example of such a stimulus is gravity, acting on rotating eggs, as described by Ancel and Vitemberger (1948) and more recently in detailed studies by Scharf and Gerhart (1980). This appears to act by causing a flow of the heavier vegetal material within the egg, initiating a different axis of bilateral symmetry. Depending on when this secondary stimulus is given, embryos can develop with either one (new) axis of bilateral symmetry, or two, resulting in twinned embryos.

Another indicator of bilateral symmetry in the fertilized egg is the plane of first cleavage, which normally divides the egg into presumptive right and left halves of the future organism. This is the first of a series of cell divisions that follow a well-defined geometrical pattern (Figure 1), subdividing the initial giant egg cell into a few thousand cells before the next stage of morphogenesis, gastrulation. This highly ordered cleavage sequence follows an orthogonal pattern of vertical followed by horizontal cleavage plans. It has been described (Goodwin and Trainor, 1980) by a series of harmonic functions, particular members of the set of possible solutions of a field equation used to describe the cleavage process. What selects this sequence of cleavage planes is the intrusic

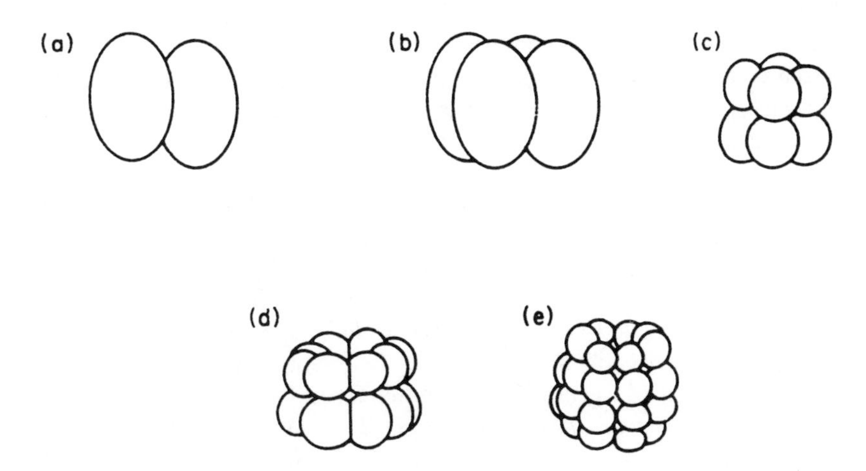


Fig. 1. The holoblastic radial cleavage pattern. (a) 2 cells, (b) 4 cells, (c) 8 cells, (d) 16 cells, (e) 32 cells (after Torrey, 1962, and Berrill and Karp, 1976).

asymmetries of the egg after fertilization, namely the animal-vegetal and the presumptive dorso-ventral axes, functioning in the model as selection rules. In addition, the cleavage sequence follows a minimum energy pathway through the set of possible transitions. However, the energy function used is not related to any experimentally measured quantity. It is a surface-energy function that is interpreted in relation to the work of cleavage, but is not rigorously derived from a consideration of the actual forces generated by the cystoskeleton and ion fluxes during this process. In relation to this description what drives the whole process, from the initial symmetry breaking of the developing oocyte to the sequence of cleavages that result in the embryonic blastula, is the intrinsic dynamics of the morphogenetic field. This is itself dependent upon metabolic energy and its conversion into the osmotic and mechanical work of morphogenesis, so that the process runs down a potential energy gradient. The 'building blocks' of the sequence are field solutions which describe the particular ways in which the laws of physics and chemistry operate to the model. Hunding (1984) has shown how Turing's type of reaction-diffusion equations, when solved for spherical geometry, can give rise to the initial patterns of the cleavage process.

There are many variations in the cleavage sequence described in Figure 1, different species showing variants with respect to rates of cell division and the exact positions of the cleavage planes, hence the size and positions of the blastomeres. Since these differences are species-specific, they are a result of the influence of gene products on the cleavage process. For example, there are species in which the cleavage planes are skewed relative to one another so that the patterns generated have a spiral form (Figure 2) rather than the vertical and horizontal order of Figure 1. This is characteristic of snails, for example, which retain a spiral form in the adult and lack bilateral symmetry. On the other hand, there are species such as certain annelids which have spiral cleavage patterns but develop bilateral symmetry in the later embryo. All of these inherited variations are a result of specific gene influences on the sequences of morphogenetic field solutions followed by a particular species in generating the adult form. Clearly there is a great diversity of sequences that can be selected or stabilized from the set of possible solutions. However, if there are constraints on morphology than there must be some constraints on the possible sequences. Let us know turn our attention to what these might be.

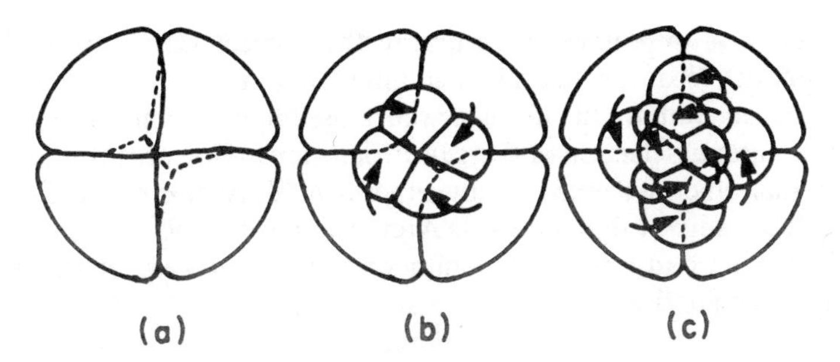


Fig. 2. The spiral cleavage pattern (dexral) in *Limnea*, seen from the animal pole. (a) 4 cells, (b) 8 cells, (c) 16 cells (after Hess, 1971).

## CONSTRAINTS ON MORPHOGENETIC SEQUENCES

I shall consider two possible sources of limitation on morphogenetic pathways, both having relevance to evolutionary and taxonomic patterns. The first of these arises from the use by Goodwin and Trainor (1980) of a variational principle to describe the set of possible solutions of the cleavage process, which implies that all the possible morphogenetic field solutions are states that minimize 'energy' according to a particular criterion. However within this set there are energy differences, and one of the selection rules specifies that at each cleavage the pattern selected is that of minimum energy among the possible alternatives. This implies that organisms move through stable generic morphological states during their development; i.e. that the forms generated are ones that arise 'naturally' by a principle of least action. Evolution is not then a process in which natural selection drives organisms into highly improbable states of increasing adaptation. Rather, ontogenies are restricted to morphogenetic patterns, such as the cleavage sequences described, which differ only slightly from one another in their energies, and it is from this limited set of possibilities that evolutionary variation arises. Morphogenetic sequences, in this view, are never displaced significantly from these generic states. This is the morphological analogue of Kauffman's and Levin's (1987) argument that organisms as genetic networks cannot be moved far from their generic states by natural selection. The implication of both propositions is that natural selection is a weak force in determining organismic states, which are limited by the intrinsic dynamic organization of the living condition.

However, it is important to recognize the conjectural nature of both propositions. What they define is an alternative research programme to that pursued within the neo-Darwinist assumption that the external force of natural selection is the major determinant of organismic state, rather than the dynamics of development. This alternative belongs within the tradition defined by D'Arcy Thompson (1919). I shall now consider a second source of limitation on morphogenetic pathways, closely related to the first.

## THE HIERARCHICAL NATURE OF MORPHOGENESIS AND TAXONOMY

One of the most obvious and basic aspects of morphogenesis is that it proceeds from the general to the particular: morphological complexity emerges gradually as progressively more localized detail emerges within an early established global order. There are essentially two different ways of viewing this process. One could regard the initial global order as providing a kind of coordinate system which specifies local domains by a positional information type of mechanism, such as different levels of concentration of 'morphogens' in monotonic gradients along different axes (Wolpert, 1971). Finer detail then emerges as genomes 'read' this information and produce specific products which guide cells into specific pathways of differentiation. Clearly the emphasis in such a scheme is on the local gene-controlled interpretation process. The alternative is to regard the global to local progression as a feature of the intrinsic dynamics of morphogenesis, initial overall order gradually giving rise to increasingly finer patterns because this is the inevitable progression of the non-linear unfolding of the field solutions. If genomes read positional information, then the detailed spatial patterns at the cellular level can emerge in one step; but if there are dynamic constraints on the order, then a hierarchical progression is inevitable.

An example that illustrates the latter type of process in a purely physical system is seen in Figure 3, which shows the development of patterns in streak lines formed in the wake of a circular cylinder in a stream of oil (from Batchelor, 1967). In hydrodynamics, the characteristics of flow fields are determined by a non-dimensional quantity, the Reynolds number ($R$). As the velocity of flow increases the Reynolds number increases and different patterns develop in the wake. At some value of $R$ between 30 and 40 the smooth stream flow becomes

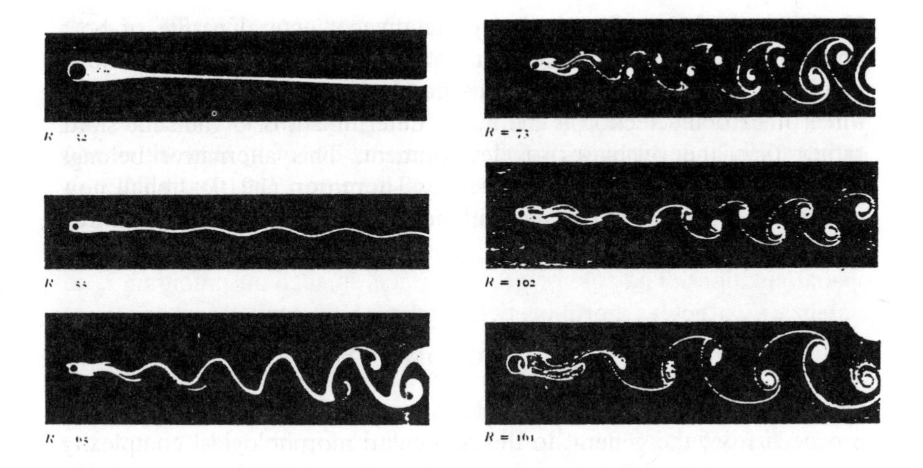


Fig. 3. Streak lines in the wake behind a circular cylinder in a stream of oil.

unstable to small disturbances and a slow oscillation arises, initially sinusoidal and with an amplitude that increases progressively downstream. This is qualitatively like the bifurcation from spatially uniform to spatially periodic concentration waves that occur in Turing diffusion-reaction fields for particular parameter values (Murray, 1977). As $R$ increases beyond this instability, more complex patterns arise with the characteristic that secondary consequences of the non-linearities of the process arise out of a primary periodicity. These develop progressively with distance downstream, hence with time from the disturbance. Thus finer and finer details emerge as the pattern unfolds from an initiating disturbance.

The point of the description is not to suggest that morphogenetic patterns originate from the hydrodynamic properties of developing organisms, though there is more to such a viewpoint than one might at first think. What I want to emphasize is simply that many pattern generating processes share with developing organisms the characteristic that spatial detail unfolds progressively simply as a result of the laws of the process. In the hydrodynamic example we see how an initially smooth fluid flow past a barrier goes through a symmetry breaking event to give a spatially periodic pattern, followed by an elaboration of local nonlinear detail which develops out of the periodicity. Embryonic development follows a similar qualitative course: initially smooth primary axes, themselves a result of spatial bifurcation from a uniform state, bifurcate to spatially periodic patterns such as segments, within which finer detail develops through cellular domains to the single cell or intracellular level by a progressive expression of nonlinearities and successive bifurcations. As this process develops, the possibilities of variation increase simply because of the combinatorics of iterated bifurcations and nonlinear detail, rising as the product of successive spatial solutions. The role of gene products in such an unfolding is to stabilize a particular morphogenetic pathway by facilitating a sequence of pattern transitions, resulting in a particular morphology. In this interpretation, gene products do not create the possibility of morphogenetic transitions, which are a result of the laws governing morphogenetic fields, just as the transitions shown in Figure 3 result from the laws of hydrodynamic flow. This elaborates further the comparison with enzymes and metabolic pathways.

There are several consequences of this view of morphogenesis. First, it is evident that morphology is generated in a hierarchical manner, from simple to complex as bifurcations result in spatially ordered asymmetries and periodicities, and nonlinearities give rise to fine local detail. Since there is a limited set of simple broken symmetries and patterns that are possible (e.g. radial, bilateral, periodic), and since developing organisms must start off laying down these elements of spatial order, it follows that these basic forms will be most common among all species. On the other hand, the finer details of pattern will be most variable between species, since the pattern generating process results in a combinatorial richness of terminal detail and specific gene products in different species stabilize trajectories leading to one or other of these. Hence, taxonomies based on morphology will inevitably be hierarchical, basic morphological patterns being common to many species while their differences are related to details of spatial patterning. In general taxonomies need not be hierarchical, as shown by the periodic table of the elements. The properties of the elements are related to both their particulars of composition in terms of neutrons, protons, and electrons (like details of species composition in relation to gene products) and to the principles of organization as described by shells or orbitals (like the principles of morphogenetic field organization), but there is no hierarchical order in terms of shared basic features and differences of detail over the elements as a whole.

Hierarchical biological taxonomies are often given an historical interpretation: the basic body plans of species are regarded as their more 'primitive' features, relating them to their common ancestors; while their more detailed characteristics, which distinguish them from others, are more recent. But any species must have both basic and detailed features. There is nothing intrinsically ancient about either. However, given that the exploration of morphogenetic potential takes place over evolutionary time scales, and that there is a much greater diversity of local detail available for evolutionary expression than of basic global patterns, it follows that the latter will have been fully explored and expressed in species early on in evolution, while exploring the enormous potential for morphological detail takes much longer, continuing to this day and beyond. The fact that virtually all the basic organismic body plans were discovered and established during an early evolutionary period, the Cambrian, is often remarked upon with surprise, but it is just what one would expect on the basis of the above argument. 'Ancient' and 'recent' morphological characters are secondary consequences of the hierarchical nature of morphogenesis and the exploration of its potential in time.

## DEVELOPMENTAL CONSEQUENCES

There are several consequences of this view of morphogenesis relating to the developmental process itself, but I shall confine myself to two. The first is the expectation that there will be many transient spatial patterns arising during the generation of embryonic structures that do not correspond to these structures themselves. For example, in the formation of the basic segmented body plan of the insects, it is to be expected that the final segmental periodicity will be preceded by longer wavelength patterns which undergo a sequence of transitions to terminal segment number. The reason for this is that fields have harmonic solutions which are usually expressed in a systematic sequence during the approach to a particular pattern; that is, the harmonic modes grow at different rates, the one that is finally stable not usually being the one that initially grows fastest. One would therefore anticipate a hierarchical sequence of spatial patterns prior to the emergence of a stable periodicity. Some mutants in *Drosophila* that affect segmentation show such a pattern, from large scale deletions (gap mutants) to 2-segment period disturbances (pair-rule mutants). Observations of the spatial distributions of pair-rule gene transcripts such as *paired* (Weir and Kornberg, 1986) reveal a further dimension of this expectation: a series of transients varying from a single period to the terminal segmented wave number (14) only one of which, the double-segment periodicity, corresponds to the mutant phenotype. This is the kind of dynamic behavior that is expected from morphogenetic fields of the type described above. It is not at all what is anticipated on the basis of a positional information type model, which need have no transients whatsoever once the initial coordinate system is set up and genomes are assigned positional values specifying how they should behave locally. The transition from global to local can occur in a single step in such a model, since the cell is the primary unit. However, this clearly does not occur. A detailed analysis of segmentation gene transcript patterns in terms of harmonic sequences and how these result in the generic properties of mutant phenotypes is given in Goodwin and Kauffman (1990) and Kauffman and Goodwin (1990).

The second consequence for morphogenesis is the expectation that a variety of perturbations should be able to override the stabilizing affects of genes on particular morphogenetic pathways, since developing organisms have a diversity of such pathways available to them in addition to the one that is normally expressed. This is precisely what happens with phenocopies. Non-specific disturbances such as ether, heat and cold shock, altered pH, etc., occurring at specific times in embryonic development result in particular alterations of body pattern, many of which mimic the effect of mutant genes but many others are as yet not duplicated mutants (i.e. genocopies have yet to be found — see Ho *et al.*, 1987). This type of behavior points strongly to law-like constraints in morphogenetic transitions (see also Goodwin, 1982, 1985, 1989) and a systematic study of these could provide a quantitative basis for a theory of the energetics of the transitions from one morphogenetic pathway to another during development.

In conclusion, it is evident that the view of morphogenesis presented in this paper has significant implications for our understanding of evolution and taxonomy. If organisms are indeed generic forms of the type described, then evolution is the exploration of the potential set of such forms as defined by morphogenetic principles. Then taxonomy derives its hierarchical nature from the intrinsic dynamics of morphogenesis; and evolution is constrained within the order defined by these dynamic principles. How much of a role natural selection then plays in this exploration remains to be established. But it is possible that it, too, is severely limited in its possible influence, apart from the necessary principle that unstable forms ('unfit' life-histories) cannot survive. Evolution would then be an historical, contingent unfolding from the potential defined by the generative dynamics of the living state, which would reveal the rational plan of organic creativity.

*Department of Biology,*
*The Open University*

## REFERENCES

Ancel, P. and Vinterberger: 1948, *Bull. Biol. Fr. Belg.* **31**, 1—182.

Batchelor, G. K.: 1967, *An Introduction to Fluid Dynamics*, Cambridge University Press.

Goodwin, B. C. and Trainor, L. E. H.: 1980, 'A field description of the cleveage process in embryogenesis', *J. Theoret. Biol.* **85**, 757—770.

Goodwin, B. C.: 1982, 'Development and Evolution', *J. Theoret. Biol.* **97**, 43—55.

Goodwin, B. C.: 1985, 'What are the causes of morphogenesis?', *Bioessays* **3**, 32—36.

Goodwin, B. C.: 1989, 'A structuralist research programme in developmental biology', in: *Dynamical structures in Biology*, B. C. Goodwin and A. Shibatani, G. Webster (eds), Edinburgh University Press.

Goodwin, B. C. and Kauffman, S. A.: 1990, 'Spatial harmonics and pattern specification in early *Drosophila* development. Part I. Bifurcation sequences and gene expression', *J. Theoret. Biol.* **144**, 303—319.

Ho, M. W., Matheson, A., Sauderns, P. T., Goodwin, B. C. and Smallcome, A.: 1987, 'Ether-induced segmentation defects in Drosophila', *Roux's Archive for Developmental Biology* (submitted).

Hunding, A.: 1984, 'Bifurcations of non-linear reaction-difussion systems in oblate spheroids', *J. Math. Biol.* **19**, 249—263.

Kauffman, S. A. and Goodwin, B. C.: 1990, 'Spatial harmonics and pattern specification in early *Drosophila* development. Part II. The four colour wheel model', *J. Theoret. Biol.* **144**, 321—345.

Kauffman, S. and Levin, S.: 1987, 'Towards a general theory of fitness on rugged landscapes', *J. Theoret. Biol.* **128**, 11—46.

Meinhardt, H.: 1982, *Models of Biological Pattern Formation*, Academic Press, London.

Murray, J. D.: 1977, *Non-Linear Differential Equation Models of Biological Systems*, Orford University Press.

Scharf, S. R. and Gerhart, J. C.: 1980, 'Determination of the dorso-ventral axis in eggs of *Xenopus laevis*: Complete rescue of UV-impaired eggs by oblique orientation before first cleavage', *Dev. Biol.* **79**, 181—198.

Thompson, D'Arcy W.: 1919, *On Growth and Form*, Cambridge: Cambridge University Press.

Turing, A. M.: 1952, 'The chemical basis of morphogenesis', *Phil. Trans. R. Soc. B* **237**, 37–72.

Wolpert, L.: 1971, 'Positional information and pattern formation', *Curr Top. Devel. Biol.* **6**, 183–224.

Weir, M. P. and Kornberg, T.: 1985, 'Patterns of *fushi tarazu* transcripts reveal novel intermediate stages in *Drosophila* segmentation', *Nature* **318**, 433–445.

SUSAN OYAMA

# IS PHYLOGENY RECAPITULATING ONTOGENY?

There are three things I would like to accomplish in these comments. First, I would like to mention some of the larger issues addressed by Stuart Kauffman's work. They are related to important questions in other fields; pointing out the relationships could be useful for finding the kinds of interdisciplinary connections to which this conference is devoted. Second, I will describe my own perspective on Kauffman's work; my interests are basically conceptual, not technical. Finally, I want to draw some parallels between arguments about phylogenetic and ontogenetic origins; these parallels cast additional light of Kauffman's endeavors.

## MAJOR ISSUES

First, then, some issues touched on by Kauffman's work. Primary, of course, is the origin of form. The question he poses is: is natural selection the only source of living form? This is a question about the role of history in evolutionary origins. This problem is related to another one: are there are fundamental differences between living and nonliving order? If natural selection is the sole source of living order, then the gap between the living and nonliving seems irreducible. If, however, biological order can be partially explained without recourse to selection, then living and nonliving systems may have important similarities. In fact, living systems become a subset of complex systems. The first two issues, then, turn on treating developmental dynamics and natural selection as alternative sources of form. Along with other critics of overly selectionist views, Kauffman also focuses on the generative role of structure. This, then, is the third major issue he addresses.

## NATURE AND NURTURE, CONSTRAINTS AND SELECTION

My perspective on these questions is a somewhat unusual one. By training, I am not an evolutionary biologist but a psychologist. In the behavioral sciences, it is not so much phylogenetic origins that are of



*Francisco J. Varela and Jean-Pierre Dupuy (eds), Understanding Origins*, 227—232.

concern, of course, as ontogenetic (developmental) ones — origins of knowledge, of body and mind, of behavior. For some years I have studied the nature–nurture opposition: its structure and its longevity (that is, its ability to persist through numerous transmutations and reformulations, despite periodic pronouncements that it is obsolete and meaningless). It has an impressive power to inform arguments in a wide range of areas, from embryology to anthropology to history. What is more to the point for this volume is that the parallels between these nature–nurture disputes and certain controversies in evolutionary studies are striking. We may ask how deep the similarities are. If they are very deep, the resolutions to both disputes may be similar as well. If (contrary to my intuition) they are not deep, the resemblances in logic and even vocabulary would be difficult to explain.

What are some similarities between the nature–nurture opposition and the controversy over evolutionary origins? For one thing, both rely on a radical separation of outsides from insides. Shaping from without is contrasted with spontaneous generation from within. In the study of ontogeny, one asks whether the world imposes itself on the body and mind or whether the organism unfolds autonomously. In phylogeny, one asks whether the environment sets problems for a species and shapes it by selection or whether order arises spontaneously. Many have noted, in fact, that psychologists' accounts of operant conditioning and biologists' accounts of evolution by natural selection invoke the same mechanisms of spontaneous variation and selection. There is even a recent rash of works exploiting this pair of mechanisms in 'cultural evolution'.

In both nature vs. nurture and constraints vs. selection we also find an opposition between fixity and malleability. In both, there is a history of oscillation between internalist and externalist views. Just as the resurgence of 'biological' views in the last few decades has challenged the long reign of behaviorism in psychology (at least American psychology), so some scholars, including Kauffman, are challenging natural selection's theoretical hegemony in biology. Amid all this shuttling between opposites, the organism is often reduced to an epiphenomenon — at best, a mere battlefield where internal and external forces contend for causal primacy, or a patchwork, fashioned partly from the inside and partly from the outside. One gains little sense of an integrated and active organism from such presentations, and the difficulties that theorists experience in striving for synthesis attests to depth and the strength of their intuitions about stasis and change, activity and passivity, order and disorder.

## A RECAPITULATION?

Is phylogeny recapitulating ontogeny? That is, are the parallels between arguments about evolutionary origins and arguments about developmental ones more than superficial likenesses? Consider the following fact: real resolution of the nature–nurture opposition can only be achieved by taking development seriously — that is, by fully acknowledging the emergence of form in ontogenetic interactions, rather than trying to attribute it to preexisting forms in the genes or in the environment. Traditional compromise solutions to the dichotomy are unsatisfactory. The most common way of compromising is to say that nature and nurture are complementary, so that some aspects of a phenotype are innate and other aspects, acquired. This produces the patchwork mentioned earlier. The many ways of distinguishing the innate from the acquired, however, are often inconsistent with each other. What is innate in the sense of being species-typical, for instance, need not be innate in the sense of being difficult to change. What is present at birth (another meaning of 'innate') is not necessarily independent of learning (still another), and so on. A somewhat more sophisticated compromise is to say that the genes determine the range of potential phenotypes, while the environment selects the specific value within that range. But the range of possible phenotypes is a function of the range of genotype–phenotype *pairings*. To attribute it only to the genes, and so to insist that 'potential' is somehow fixed at conception, is to engage in a modern version of preformationism. (For fuller discussion of these attempted solutions, see Oyama, 1989.)

The same attempts at compromise appear in the evolutionary literature. In a review of books on development and evolution, for example, Keith Thomson speaks of externalist and internalist views of causality in evolution and asserts that they should be seen as complementary (1985). Kauffman (this volume) advocates something similar. The range-of-potential argument is also found in the literature on developmental constraints, but this time potential is defined by developmental dynamics while the genes select specific solutions (Alberch, 1982; Webster and Goodwin, 1982).

I said that synthesis of the nature–nurture opposition requires taking development seriously. Ontogeny is neither the revelation of an internal essence nor the imposition of an external order. 'Nature' is the emergent phenotype, constructed in development through the interactions of a highly ramified organism–environment system (Oyama, 1985). Is it possible that an analogous synthesis could be achieved for evolutionary origins as well? Could it be that the resolution of the oppositions between internal and external forces in phylogeny will recapitulate the one I have described for the study of ontogeny (Gray, 1987)? Since development is not autonomously driven from inside, but is the result (and cause) of multilevelled interaction, should the internality and necessity of developmental constraints on selection be reconsidered? And since selection is not the action of an external agent but a shorthand term for the results of certain kinds of interactions within organisms (and between organisms and their surround), should the externality and arbitrariness of natural selection be reconsidered as well?

I believe this is the case. If I am correct, then the way to counteract excessive claims about the power of selection is not to oppose them by invoking internal constraints but to reconceptualize the process in a way that eliminates the opposition altogether. If, on the other hand, the two oppositions are very different, how do we explain the unnerving parallels between them? The similarities exist at the levels of vocabulary, conceptual structure, and even historical dynamics.

## MORE PLAYERS OR A DIFFERENT GAME?

While I appreciate the richness of Kauffman's critique of selectionism, his investigations of self-organization and the emergence of higher-level order, and while I agree that conventional evolutionary theory should be broadened, I have some misgivings about his approach. Long experience with the nature–nurture complex has made me wary of some of the arguments I find in evolutionary studies. I suspect that, as in the nature–nurture complex, it is not enough to oppose external formation with internal. We must alter the very ground on which opponents take their positions. As pleased as I am to see orthodox, externalist theory challenged, it is a bit like seeing behaviorism challenged by some sociobiologists and other proponents of neoinstinctivist, 'biological' approaches. It allows more players to join the game, but the game itself remains unchanged.

If my recapitulationist thesis is valid, it may be better to challenge the very conceptual framework that demands such oppositions in the first place. Development should be at the heart of evolutionary theory. We have seen that proper appreciation of its processes allows us to transcend the age-old dichotomy between nature and nurture; it may be that serious consideration of development could help us resolve the opposition between internal constraints and selection as well. Developmental studies have been virtually excluded from the neo-Darwinian synthesis, even though it is developmental processes that generate the series of interacting life cycles which constitute evolution.

Kauffman's description of cell types, and especially of the 'poised' readiness of cells to differentiate, are welcome acknowledgments of ontogenetic processes. When I look at his computer simulations, however, I generally find it hard to see phenotypes emerging in development. Perhaps the inclusion of interactive, multileveled processes of development in the simulations might help bridge the gap between genetic regulatory circuits and organisms, between internal constraints and external pressures, between an historical necessity and historical contingency. I suspect that the more development is included, the harder it will be to calculate fitness by counting genes, for example, and that the more the organism is placed in its developmental and ecological contexts, the harder it will be to think of fitness as a property of individual organisms at all. Such an integration of ontogeny into the phylogenetic story could, I think, help us attain deeper understanding of the origins of living order by synthesizing processes that have traditionally been treated as alternatives. Whether or not this is the case, we should be reflective about the kinds of arguments we enter and the assumptions that underlie them.

*CUNY, New York*

## REFERENCES

Alberch, P.: 1982, 'Developmental constraints in evolutionary processes', in: Bonner, J. T. (Ed.), *Evolution and Development*. Berlin: Springer-Verlag, pp. 313—332.

Gray, R.: 1986, 'Beyond labels and binary oppositions: what can be learnt from the nature/nurture dispute?' *Rivista di Biologia* **80**, 192—196.

Oyama, S.: 1985, *The Ontogeny of Information: Developmental Systems and Evolution*. Cambridge: Cambridge University Press.

Oyama, S.: 1989, 'Ontogeny and the central dogma: do developmentalists need the concept of genetic programming in order to have an evolutionary perspective?' in: Gunnar, M. (Ed.), *Minnesota Symposium on Child Psychology, Vol.* **22**.

Thomson, K. S. : 1985, 'Essay review: the relationship between development and evolution', *Oxford Surveys in Evolutionary Biology* **2**, 220—233.

Webster, G. and Goodwin, B. C.: 1982, 'The origin of species: a structuralist approach', *Journal of Social and Biological Structures* **5**, 15—47.

# PART IV

# PERCEPTION AND THE ORIGIN OF COGNITION

FRANCISCO J. VARELA

# WHENCE PERCEPTUAL MEANING? A CARTOGRAPHY OF CURRENT IDEAS

## 1. INTRODUCTION

### *Clarifications*

This essay was written for the purpose of providing a minimal common ground for discussion. It is, of necessity, an ambitious attempt to give a concise account of the various current ideas on the origin of meaning in living and artificial systems in such a way that it is accessible to an interdisciplinary audience, and yet substantive enough to produce debate among the specialists. I apologize at the outset to both groups for passages that will seem irritatingly simple or too abstruse.

Also I have restrained myself to basic or 'lower' cognitive abilities, that is, issues closer to perception, motion, and simple learning. This is in contrast to 'higher' cognitive abilities, issues closer to language and reasoning.

This presentation cannot be neutral and my preferences are quite explicitly laid out in the text. In particular, in this essay I will argue that the kingpin of cognition is its capacity for *bringing forth* meaning: information is not pre-established as a given order, but regularities emerge from a co-determination of the cognitive activities themselves.

### *Outline*

Cognitive Science (CS) is a little over 40 years old. It is *not* established as a mature science with a clear sense of direction and a large number of researchers constituting a community, as is the case of, say, atomic physics or molecular biology. Accordingly, the future development of CS is far from clear, but what has already been produced has had a profound impact, and this will continue to be the case. But progress in the field is based on daring conceptual bets (somewhat like trying to put a man on the moon . . . without knowing where the moon is). For the sake of concretness, the federation of disciplines I take here as forming cognitive science today are neuroscience, artificial intelligence, cognitive psychology, linguistics, and epistemology.



*Francisco J. Varela and Jean-Pierre Dupuy (eds), Understanding Origins*, 235—263.

The main purpose of this background paper is to provide an X-ray picture of the current state of affairs of CS, in regards to perception and the origin of meaning. Now, like anybody who has ever examined a scientific discipline with any proximity, I have found the cognitive sciences to be a *diversity* of semi-compatible visions, and not a monolithic field. Further, as any social activity, it has poles of domination so that some of its participating voices acquire more force than others at different periods of time. This is strikingly so in the modern cognitive science revolution which was heavily influenced through some lines of research, particularly in the USA. My bias here is to emphasize diversity.

I will proceed in four stages which are conceptually and practically quite distinct. These four stages are the following:

- Stage 1: A glance at the foundational years (1943—1953);
- Stage 2: Symbols: The cognitivist paradigm;
- Stage 3: Emergence: alternatives to symbol manipulation;
- Stage 4: Enaction: alternatives to representations.

Through this four-tiered description and their articulations we will examine the basis of what is already established as a clear trend (Stages 1 and 2), and the fact that this established paradigm coexists with a wider spectrum of perspectives (Stages 3 and 4). This challenging heterodoxy has the potential for deep changes.

## 2. A GLANCE AT THE FOUNDATIONAL YEARS

We start with a brief look into the roots of these ideas in the decade 1943—1953, so as to touch on the issues of direct relevance for us here[1]. In fact, virtually all the themes in active debate today were already introduced in these formative years, evidence that they are deep and hard to tackle. The 'founding fathers' knew very well that their concerns amounted to a new science, and christened it with a new name: *cybernetics*. This name is not in current use any more, and many cognitive scientists today would not even recognize the family connection. This is not idle. It reflects the fact that to become established as a science, in its clear-cut cognitivist orientation (Stage 2 in this text), the future cognitive science had to sever itself from its roots; more complex and fuzzier, but also richer. This is often the case in the history of science: it is the price of passing from an exploratory stage to becoming a research program — from cloud to crystal.

### *The Fruits of the Cybernetics Movement*

The cybernetics phase of CS produced an amazing array of concrete results, apart from its long-term (often underground) influence. Some of these are:

- the use of mathematical logic to understand the operation of the nervous system;
- the invention of information processing machines (as digital computers), thus laying the basis for artificial intelligence;
- the establishment of the metadiscipline of system theory, which has had an imprint in many branches of science, such as engineering (system analysis, control theory), biology (regulatory physiology, ecology), social sciences (family therapy, structural anthropology management, urban studies), and economics (game theory);
- information theory as a statistical theory of signal and communication channels;
- the first examples of self-organizing systems.

The list is impressive: we tend to consider many of these notions and tools as an integral part of our lives. Yet they were all inexistent before this formative decade, and they were all produced by intense exchange among people of widely different backgrounds: a uniquely successful interdisciplinary effort.

### *Logic and the Science of Mind*

The avowed intention of the cybernetics movement was to create a *science of mind*. To its readers, the mental phenomena had been for far too long in the hands of psychologists and philosophers, and they felt themselves called to express the processes underlying mental phenomena in explicit mechanisms and mathematical formalisms.[2]

One of the best illustrations of this mode of thinking (and its tangible consequences) was the seminal: 'A logical calculus immanent in nervous activity'[3] (1943), paper by McCulloch and Pitts. Several major leaps were taken in this article. First, proposing that *logic* is the proper discipline with which to understand the brain and mental activity. Second, seeing the brain as a device which *embodies* logical principles in its component elements of neurons. Each neuron was seen as a threshold device being either active or inactive. Such simple neurons could then be connected to one another, their interconnections performing the role of logical operations so the entire brain could be regarded as a deductive machine.

These ideas were central for the invention of the digital computers.[4] At that time, vacuum tubes were used to implement the McCulloch-Pitts neurons, where today we find silicon chips, but modern computers are still built on the same von Neumann architecture. This major technological breakthrough also laid the basis for the dominant approach to the scientific study of mind that was to crystallize in the next decade as the cognitivist paradigm.

### *The End of an Era*

There was, of course, a lot more to this creative decade. For instance, the debate about whether logic was indeed sufficient to understand the brain, because it neglected its distributed qualities. Alternative models and theories were put forth, which for the most part were to lay dormant until revived to constitute an important alternative for CS in the 1970s (Stage 3). By 1953, in contrast to their initial vitality and unity, the main actors of the cybernetics phase were distanced from each other and many died shortly thereafter. The idea of mind as logical calculation was to be continued.

## 3. SYMBOLS: THE COGNITIVISTS HYPOTHESIS

### *Enter the Cognitivists*

Just as 1943 was clearly the year in which the cybernetics phase was born, so was 1956 clearly the year which gave birth to the second phase of CS. During this year, at two meetings held at Cambridge and Dartmouth, new voices (like those of Herbert Simon, Noam Chomsky, Marvin Minsky and John McCarthy) put forth ideas which were to become the major guidelines for modern cognitive science.[5]

The central intuition is that intelligence (including human intelligence) so resembles a computer in its essential characteristics that cognition can be *defined* as computations of symbolic representations. Clearly this orientation could not have emerged without the basis laid during the previous decade, and evoked in the previous section. The main difference is that one of the many original tentative idea is promoted here to full blown *hypothesis*, with a strong desire to set its boundaries apart from its broader, exploratory, and interdisciplinary roots where the social and biological sciences figured pre-eminently with all their multifarious complexity. *Cognitivism*[6] is a convenient label for this large but well-delineated orientation, that has motivated many scientific and technological developments since 1956, in all the areas of cognitive science.

### *An Outline of the Doctrine*

The cognitivist research programme can be summarized as answers to the following questions:

> *Question # 1*: What is cognition:
> *Answer*: Information processing: Rule-based manipulation of symbols.
>
> *Question # 2*: How does it work?
> *Answer*: Through any device which can support and manipulate discrete physical elements: the symbols. The system interacts only with the form of the symbols (their physical attributes), not their meaning.
>
> *Question # 3*: How do I know when such a cognitive system is functioning adequately?
> *Answer*: When the symbols appropriately represent some aspect of the real world, and the information processing leads to a successful solution of the problem posed to the system.

Obviously the cognitivist programme as outlined above did not come out ready-made, like Athena from the head of Zeus. We are presenting it with the benefits of 30 years of hindsight. However, not only has this bold research programme become fully established, but even today is *identified* by many with cognitive science itself, although this is changing rapidly. Until very recently, only a few among its active participants, let alone in the public at large, were sensitive to its roots or its current challenges and alternatives. "The brain processes information from the outside world" is a household phrase understood by everybody. It is odd to treat statements such as this as problematic rather than obvious, and the ensuing conversation will immediately be labeled as being 'philosophical'. This is a *blindness* in contemporary common sense introduced in our culture after the establishment of cognitivism.

### *What Cognitivism has Wrought: Artificial Intelligence*

The manifestations of cognitivism are nowhere more visible than in artificial intelligence, which is the *literal construal* of the cognitivist hypothesis. Over the years many interesting theoretical advances and technological applications have been made within this orientation: expert systems, robotics, image processing. These results have been widely publicized, and we need not insist on examples here.

Because of its wider implication, however, it is worth noting that AI and its cognitivist basis has reached a dramatic climax in Japan's ICOT Fifth Generation Program. For the first time since the war there is a national plan concerting the efforts of industry, government, and universities. The core of this program — the rocket to be put on the moon by 1992 — is a cognitive device capable of understanding human language, and of writing its own programs when presented with a task by an untrained user. Not surprisingly, the heart of the ICOT program is the development of a series of interfaces of knowledge representation and problem solving based on PROLOG, a high level programming language for predicate logic. The ICOT program has triggered immediate responses from Europe and in the USA, and there is little question that this is a major commercial and engineering battlefield. However, what concerns us here is not whether the rocket will be built or not, but whether it points where the moon is. More about this latter.

### *Cognitive Psychology*

The cognitivist hypothesis finds its most literal construal in AI. Its complementary endeavour is the study of natural, biologically implemented cognitive systems, most especially man. Here, too, computationally characterizable representations have been the main explanatory tool. Mental representations are taken to be occurrences of a formal system, and the mind's activity is what gives these representations their attitudinal colour: beliefs, desires, plans, and so on. Here, therefore, unlike AI, we find an interest in what the *natural* cognitive systems are really like, and it is assumed that their cognitive representations are *about* something *for* the system, they are intentional.[7]

A good example of this orientation of research is the following. Subjects were presented with geometric figures and asked to rotate them in their heads. They consistently reported that the difficulty of the task depended on the number of degrees of freedom in which the figure had to be rotated. That is, everything happens as though we have a mental space where figures are rotated like on a television screen.[8] In due time these experiments produced an explicit theory postulating rules by which the mental space operates, similar to those used on computer displays operating on stored data. These researchers proposed that there is an interaction between language-like operations and picture-like operations, and together they generate our internal eye.[9] This approach has generated an abundant literature, both for and against,[10] and every level of the observations has been given alternative interpretations. However, the study of imagery is a perfect example of the way the cognitivist approach proceeds when studying mental phenomena.

### *Information Processing in the Brain*

Another equally important effect of cognitivism is the way it has shaped current views about the brain. Over the years almost all of neurobiology (and its huge body of empirical evidence) has become permeated with the information-processing perspective. More often than not, the origins and assumptions of this perspective are not even questioned.[11]

The best example of this approach is given by the celebrated studies on the visual cortex, where one can detect electrical responses from neurons when the animal is presented with a visual image. It was reported early on that it was possible to classify these cortical neurons as 'feature' detectors, responding to certain attributes of the object being presented: its orientation, contrast, velocity, colour, and so on.[12] In line with the cognitivist hypothesis, these results were seen as giving biological substance to the notion that the brain picks up visual information from the retina through the feature specific neurons in the cortex, and the information is then passed on to later stages in the brain for further processing (conceptual categorization, memory associations, and eventually action).

In its most extreme form, this view of the brain is expressed in Barlow's[13] grandmother cell doctrine, where there is a correspondence between concepts or percepts and neurons. (This is the AI equivalent of detectors and labeled lines.)

### *A Brief Outline of Dissent*

CS-as-cognitivism is a well-defined research programme, complete with prestigious institutions, journals, applied technology and international commercial concerns. Most of the people who work within CS would subscribe — knowingly or unknowingly — to cognitivism or its close variants. After all, if one's bread and butter consists in writing programs for knowledge representation, or finding neurons for well-defined tasks, how could it be otherwise? For our concerns here, it is important to draw attention to the depth of this *social commitment* from a large sector of the research community in CS. We now focus on the dissent, taking two basic forms:

- A critique of symbolic computations as the appropriate carrier for representations;
- A critique of adequacy of the notion of representations as the Archimides's point for CS.

## 4. EMERGENCE: ALTERNATIVES TO SYMBOLS

### *The Roots of Self-Organization Ideas*

Alternatives to the towering dominance of logic as the main approach to CS had already been proposed and widely discussed during the formative decade. At the Macy Conferences, for example, it was argued that in actual brains there are no rules or central logical processor nor is information stored in precise addresses. Rather, brains seem to operate on the basis of massive interconnections, in a distributed form, so that their actual connectivity changes as a result of experience. In brief, they present a self-organizing capacity that is nowhere to be found in logic. In 1958 F. Rosenblatt built the 'Perceptron', a simple device with some capacity for recognition, purely on the basis of the changes of connectivity among neuron-like components;[14] similarly, W. R. Ashby carried out the first study of the dynamics of very large systems with random interconnections, showing that they exhibit coherent global behaviors.[15]

History would have it that these alternative views were bracketed out of the intellectual scene in favour of the computational ideas discussed above. It was only during the late 70s that an explosive rekindling of these ideas took place — after 30 years of preeminence of the cognitivist orthodoxy; what D. Dennett[16] has called High Church Computationalism. Certainly one of the contributing factors for this renewed interest was the parallel rediscovery of self-organizational ideas in physics and non-linear mathematics.

### *Motivation to Look for an Alternative*

The motivation to take a second look at self-organization was based on two widely acknowledged deficiencies of cognitivism. The first is that symbolic information processing is based on *sequential* rules, applied one at the time. This famous von Neumann bottleneck is a dramatic limitation when the task at hand requires large numbers of sequential operations (such as natural image analysis or weather forecasting). A continued search for *parallel* processing algorithms on classical architectures has met with little success because the entire computational philosophy runs precisely counter to it.

The second important limitation is that symbolic processing is *localized*: the loss of any part of the symbols or rules of the system implies a serious malfunction. In contrast a *distributed* operation is highly desirable, so that there is at least a relative equipotentiality and immunity to mutilations.

These two deviations from cognitivism can be phrased as the same: the architectures and mechanisms are far from biology. The most ordinary visual tasks, done even by tiny insects, are done faster than is physically possible when simulated in a sequential manner; the resiliency of the brain to damage without compromising all of its competence, has been known to neurobiologists for a long time.

### *What is Emergence?*

The above suggests that instead of focusing on symbols as a starting point, one could start with simple (non-cognitive) components which would connect to one another in dense ways. In this approach each component operates only in its *local* environment, but because of the network quality of the entire system, there is global cooperation which *emerges* spontaneously, when the states of all participating components reach a mutually satisfactory state, without the need for a central processing unit to guide the entire operation.[17] This passage from local rules to global coherence is the heart of what used to be called self-organization during the foundations years.[18] Today, different people prefer to speak of emergent or global properties, network dynamics, or even synergetics. Although there is no unified formal theory of emergent properties, the most obvious regional theory is that of attractors in dynamical systems theory.[19] These are not the property of an individual components, but of the entire system, yet each component contributes to its emergence and characteristics.

### *A Change of Perspective Concerning the Brain*

Recent work has produced some detailed evidence of how emergent properties are at the core of the brain's operation. This is hardly surprising if one looks at the details of the brain's anatomy. For example, although neurons in the visual cortex do have distinct responses to specific 'features' of the visual stimuli, as mentioned above, this is valid in an anesthetized animal with a highly simplified (internal and external) environment. When more normal sensory surroundings are allowed, and the animal is studied awake and behaving, it has been shown that and the stereotyped neuronal responses previously described become highly context sensitive. For example, there are distinct effects produced by bodily tilt[20] or auditory stimulation.[21] Further, the neuronal response characteristics depend directly on neurons localized far from their receptive fields.[22]

Thus, it has become increasingly necessary to study neurons as members of *large ensembles* which are constantly disappearing and arising through their cooperative interactions, and where every neuron has multiple and changing responsiveness to visual stimulation, depending on context. Even at the most peripheral end of the visual system, the influences that the brain receives from the eye is met by more activity that *descends* from the cortex. It is by the encounter of these two ensembles of neuronal activity that a new coherent configuration emerges, depending on the match : mismatch between the sensory activity and the 'internal' setting at the cortex.[23] In general, an individual neuron participates in many such global patterns and bears little significance when taken individually.

Although these examples are taken from the domain of vision for the sake of contrast with the example of the previous section, several other detailed analysis have proliferated recently.[24] We do not need to insist on this point further.

### *The (Neo-)Connectionist Strategy*

The brain has been (once more) a main source of metaphors and ideas for other fields of CS in this alternative orientation. Instead of starting from abstract symbolic descriptions, one starts with a whole army of simple stupid components, which, appropriately connected, can have interesting global properties. These global properties are the ones that embody/express the cognitive capacities being sought.

The entire approach depends, then, on the introduction of the appropriate connections and this is usually done through a rule for gradual *change* of connections starting from a fairly arbitrary initial state. Several such rules are available today, but by far the most explored is Hebb's Rule, whereby changes in connectivity in the brain could arise from the degree of *coordinated* activity between neurons: if two neurons tend to be active together, their connection is strengthened; otherwise it is diminished. Therefore the system's connectivity becomes inseparable *from its history of transformation*, and related to the kind of task defined for the system. Since the real action happens at the level of the connections, the name (neo)*connectionism* has been proposed for this direction of research.[25]

One of the important factors for the explosive interests in this approach today was the introduction of some effective methods to follow network changes, most notably statistical measures which provide the system with a global 'energy' function that assures its convergence.[26] For instance, take $N$ simple neuron-like elements, connect them reciprocally, and provide them with a Hebb-type rule. Next present this system with a succession of (non-correlated) patterns at some of its nodes, and at each presentation let the system reorganize itself by rearranging its connections following its energy gradient. After the learning phase, when the system is presented again with one of these patterns, it recognizes it, in the sense that it falls into a unique attractor, and internal configuration that is said to represent the learned item. The recognition is possible provided the number of patterns presented is not larger than about $0.15N$. Furthermore, the system performs a correct recognition even if the pattern is presented with added noise, or the system is partially mutilated.[27]

Another important technique favoured by some researchers is back-propagation: changes in neuronal connections inside the network (hidden units) are assigned so as to minimize the difference between the network's response and what is expected of it, much like somebody trying to imitate an instructor.[28] NetTalk, a celebrated recent example of this method, is a grapheme-phoneme conversion machine that works by being shown a few pages of English text in its learning phase. As a result, NetTalk can read out loud a new text, in what many listeners consider deficient but comprehensible English.[29]

Connectionist models provide, with amazing grace, a working model for a number of basic cognitive capacities, such as rapid recognition, associative memory and categorical generalization. The current work with this orientation is justified on several counts. First, cognitivist AI and neuroscience had few convincing results to account for or reconstruct some of the cognitive performances just described. Second, these models are quite close to biological systems, and this means that one can work with a degree of integration between AI and neuroscience that was hitherto unthinkable. Finally, the models are general enough to be applied, with little modification, to a variety of domains.

### *An Outline of the Doctrine*

This alternative orientation — connectionist, emergent, self-organization, associationist, network dynamical — is young and diverse. Most of those who would enlist themselves as members hold widely divergent views on what CS is and on its future. Keeping this disclaimer in mind here are the alternative answers to the previous questions:

> *Question # 1*: What is cognition?
> *Answer*: The emergence of global states in a network of simple components.
>
> *Question # 2*: How does it work?
> *Answer*: Through local rules for individual operation, and rules for changes in connectivity between the elements.
>
> *Question # 3*: How do I know when a cognitive system is functioning adequately?

*Answer*: When the emergent properties (and resulting structure) can be seen to correspond to a specific cognitive capacity: a successful solution of a required task.

### *Exeunt the Symbols*

One of the most interesting aspects of this alternative approach to CS in that symbols, in their conventional sense, play no role. This entails a radical departure from a basic cognitivist principle: the physical structure of symbols, their form, is forever separated from what they stand for, their meaning. This separation between form and meaning was the master stroke that created the computational approach, but it also implies a weakness when addressing cognitive phenomena at a deeper level. How do symbols *acquire* their meaning? Whence this extra activity which is, by construction, not in the cognitive system?

In situations where the universe of possible items to be represented is constrained and clear-cut (such as when a computer is being programmed, or when an experiment is conducted with a set of predefined visual stimuli), the assignment of meaning is clear. Each discrete physical item within the cognitive system is made to correspond to an external item (its referential meaning), a mapping operation which the observer easily provides. Remove these constraints, and the form of the symbol is *all* that is left, and meaning becomes a ghost, as it would if we were to contemplate the bit patterns in a computer whose operating manual was lost.

In the connectionist approach, meaning is linked to the overall performance (say in recognition or learning). Hence, meaning relates to the global state of the system, and is not located in a particular symbols. The form/meaning distinction at the symbolic level disappears, and reappears in a different garb: the observer provides the correspondence between the system's global state and the world it is supposed to handle. This, is, then, a radically different way of working with representations. We shall return to this issue below.

## 5. LINKING SYMBOLS AND EMERGENCE

At this stage the obvious question to consider is the *relation* between the symbolic and emergent views on the origin of simple cognitive properties. The obvious answer is that these two views should be complementary top-down and bottom-up approaches, or that they should be pragmatically adjoined in a some mixed mode, or simply used at different stages. A typical example of this move is to describe early vision in connectionist terms, say up to primary visual cortex, but to assume that at the inferotemporal cortex level, the description should be based on symbolic programs. But the conceptual status of such synthesis is far from clear, and concrete examples are still lacking.

In my view the most interesting relation between emergent and symbolic descriptions is one of *inclusion*, that is, the view of symbols as a higher level of description of properties embedded in an underlying distributed system. The case of the so-called genetic code is paradigmatic, and I will use it here for concreteness. For many years biologists considered protein sequences as being instructions coded in DNA. However, it is clear that DNA triplets are capable of predictably specifying an aminoacid in a protein if and only if they are embedded in the cell's metabolism, that is, in the midst of thousands of enzymatic regulations in a complex chemical network. It is only by the emergent regularities of such network as a whole that we can bracket out this metabolic background, and treat triplets as codes for aminoacids. In other words, the symbolic description is possible at another level of description. Clearly, it is possible to treat such symbolic regularities in their own right, but their status and interpretation is quite different than if taken at face value, with independence of the substratum from which they arise.[30]

The example of genetic information can be transposed directly to the cognitive networks with which neuroscientists and connectionist deal. In fact, some researchers have recently expressed this point of view.[31] In Smolesnky's harmony theory[32] for example, fragmentary atoms of 'knowledge' about electrical circuits linked by distributed statistical algorithms, yields a model of intuitive reasoning in this domain. The competence of this whole system can be described as doing inferences based on symbolic laws, but their performance sits at a different level and is never achieved by reference to a symbolic interpreter. This point is graphically portrayed in Figure 1.

Thus, a fruitful link between a less orthodox cognitivism, relaxed to emerge from parallel distributed processing provided by self-organizational approaches is a concrete possibility, especially in engineering-oriented AI. This potential complementation will undoubtedly produce visible results, and might well become the dominant trend for many years in CS.

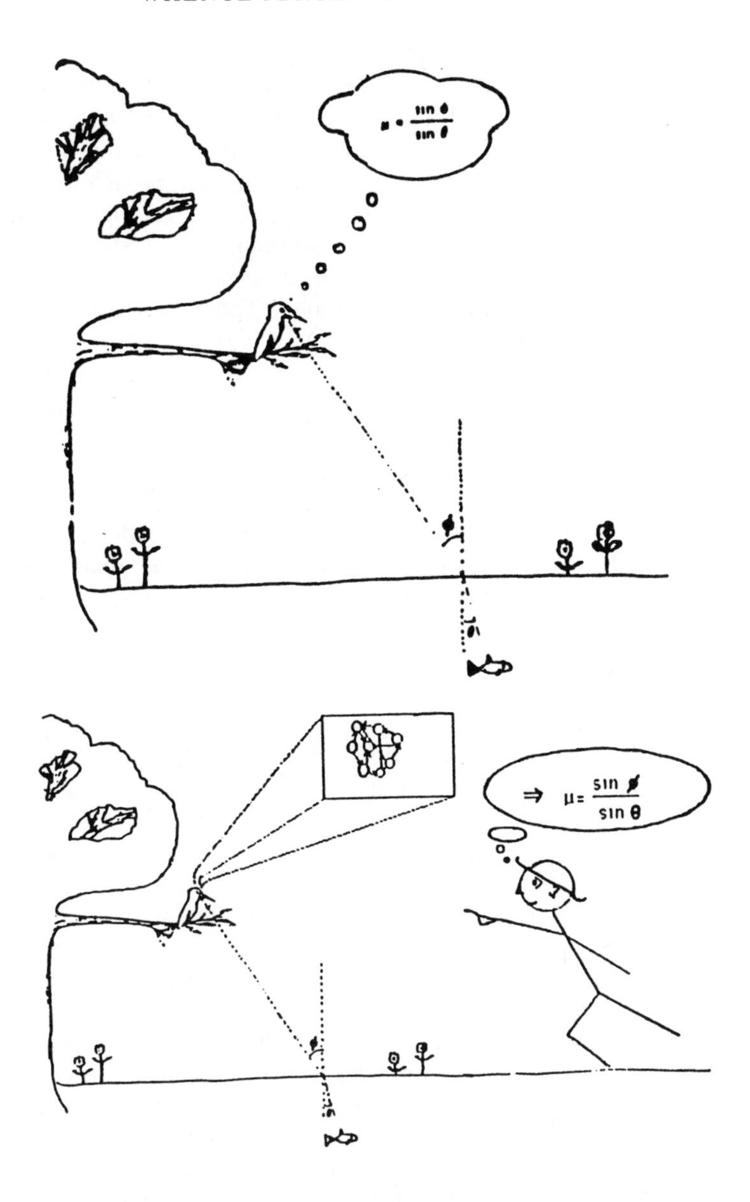


Fig. 1. (a) A *Punch* cartoon that depicts succinctly the cognitivist hypothesis. To catch its prey, this kingfisher has, in its brain, a representation of Snell's law of refraction. (b) Another reading of the cartoon to indicate how the symbolic levels can be seen as arising from the underlying network.

This move is, of course, inadmissible from a strict or orthodox cognitivist position.[33] Among the many issues that change from the emergence viewpoint, two of them are worth underlining here. First, the question of the origin of a symbol and its signification (i.e. why does ATT code for alanine?) has at least a clear way to approach it. Second, any symbolic level becomes highly dependent on the underlying network's properties and peculiarities, and bound to is history. A purely procedural account of cognition, independent of its embodiments and history, is, therefore, seriously questioned. These two issues take us straight into our last stage.

## 6. ENACTION: ALTERNATIVES TO REPRESENTATIONS

### *Further Grounds for Dissatisfaction*

It is tempting to stop in an analysis of today's CS with just the two approaches already discussed. But this would be inadequate, since in both orientations (and hence some future synthesis) some essential dimensions of cognition would be *still* missing. We need to keep in mind a larger horizon for CS, born from a deeper dissatisfaction than the search for alternatives to symbols, and closer to the very foundation of representational systems. Hopefully this orientation, enjoying today some breathing space, will not suffer the same fate as that of earlier self-organization ideas, left to be rediscovered after 30 years.

### *Insisting on Common Sense*

The central dissatisfaction of what we here call the enactive alternative is simply the complete absence of common sense in the definition of cognition so far. Both in cognitivism (by its very basis) and in present day connectionism (by the way it is practised), it is still the case that the criteria for cognition is a successful representation of an external world which is pre-given, usually as a problem solving situation. However, our knowledge activity in everyday life reveals that this view of cognition is too incomplete. Precisely the greatest ability of all living cognition is, within broad limits, to *pose* the relevant issues to be addressed at each moment of our life. They are not pre-given, but *enacted or brought forth* from a background, and what counts as relevant is what our common sense sanctions as such, always in a contextual way.

This is a critique of the use of the notion of representation as the core of CS, since only if there is a pre-given world can it be represented. If the world we live in is brought forth rather than pre-given, the notion of representation cannot have a central role any longer. The depth of the assumptions we are touching here should not be underestimated, since our rationalist tradition as a whole has favoured (with variants of course) the understanding of knowledge as a mirror of nature. It is only in the work of some continental thinkers (most notably M. Heidegger, M. Merleau-Ponty and M. Foucault) that the explicit critique of representations, and the enactive dimension of understanding has begun. These hermeneutical themes were first introduced as the discipline of the interpretation of ancient texts, but has now been extended to denote the entire phenomenon of *interpretation* understood as the activity of enactment or bringing forth, to which we are alluding.[34] Since we are concerned here with the dominance of usage, instead of with representations, it is appropriate to call this alternative approach to CS *enactive*.[35]

In recent years, however, a few researchers within CS have put forth concrete proposals, taking this critique from the philosophical level into the laboratory and into specific work in AI. This is a more radical departure from CS than the preceding one, and one that goes beyond the themes discussed during the formative period. At the same time, it naturally incorporates the ideas and methods developed within the connectionist context, as we shall presently see.

### *The Problem with Problem Solving*

The assumption in CS has all along been that the world can be divided into region of discrete elements and tasks to which the cognitive system addresses itself, acting within a given 'domain' of problems: vision, language, movement. Although it is relatively easy to define all possible states in the 'domain' of the game of chess, it has proven less productive to carry this approach over into, say, the 'domain' of mobile robots. Of course, here too one can single out discrete items (such as steel frames, wheels and windows in a car assembly). But it is also clear that while the chess world ends neatly at some point, the world of movement amongst objects does not. It requires our continuous use of common sense to configure our world of objects.

In fact, what is interesting about common sense is that it cannot be packaged into knowledge at all, since it is rather a readiness-to-hand or know-how based on lived experience and a vast number of cases, which entails an embodied history. A careful examination of skill acquisition, for example, seems to confirm this point.[36] A lived, natural world does not have sharp boundaries, and thus we expect a symbolic representation with rules, to be unable to capture common-sense understanding. In fact, it is fair to say that by the 1970s, after two decades of humblingly slow progress, it dawned on many workers in CS that even the simplest cognitive action requires a seemingly infinite amount of knowledge, which we take for granted (in fact it is so obvious as to be invisible), but which must be spoon-fed to the computer. The cognitivist hope for a general problem-solver in the early 60s, had to be shrunk down to local knowledge domains with well-posed problems to be solved, where the programmer could project onto the machine as much of his/her own background knowledge as was practicable. Similarly, the commonly practiced connectionist strategy depends on restricting the space of possible attractors by means of assumptions about known properties of the world which are incorporated as additional constraints for regularization,[37] or, more recently, in back propagation methods as a perfect model to be imitated. In both instances, the unmanageable ambiguity of background common sense is left at the periphery of the inquiry, hoping that it will be clarified in due time.[38]

Such acknowledged concerns have a well-developed philosophical counterpart. Phenomenologists of the continental tradition have produced detailed discussions as to why knowledge is a matter of being in a world which is inseparable from our bodies, our language and social history.[39] It is an ongoing interpretation which cannot be adequately captured as a set of rules and assumptions since it is a matter of action and history, an understanding picked up by imitation and by becoming a member of an understanding which is already there. Furthermore, we cannot stand outside the world in which we find ourselves, to consider how its contents match their representations of it: we are always already immersed in it. Positing rules as mental activity is factoring out the very hinge upon which the living quality of cognition arises. It can only be done within a very limited context where almost everything is left constant, a pervasive *ceteris paribus* condition. Context and common sense are not residual artifacts that can be progressively eliminated by the discovery of more sophisticated rules. They are in fact the very essence of *creative* cognition.

If this critique is correct, even to some limited degree, progress in understanding cognition as it functions normally (and not exclusively in highly constrained environments) will not be forthcoming unless we are to start from another basis than a domain out-there to be represented.

### *Exeunt the Representations*

The real challenge posed to CS by this orientation is, then, that it brings into question the most entrenched assumption of our scientific tradition altogether: that the world as we experience it is independent of the knower. Instead, if we are forced to conclude that cognition cannot be properly understood without common sense, and this is none other than our *bodily and social history*, the inevitable conclusion is that knower and known, subject and object, stand in relation to each other as mutual specification: they arise together.

Consider the case of vision: which came first, the world or the image? The answer of vision research (both cognitivist and connectionist) is unambiguously given by the names of the tasks investigated: to 'recover shape from shading' or 'depth from motion', or 'colour from varying illuminants'. This we may call the *chicken* extreme:

- *Chicken* position: The world out-there has fixed laws, it precedes the image that it casts on the cognitive system, whose task is to capture it appropriately (whether in symbols or in emergent states).

Now, notice how very reasonable this sounds, and how difficult it seems to imagine that it could be otherwise. We tend to think that the only alternative is the *egg* position:

- *Egg* position: The cognitive system creates its own world, and all its apparent solidity is the primary reflection of the internal laws of the organism.

The enactive orientation proposes that we take a middle way,[40] moving beyond these two extremes by realizing that (as farmers know) egg and chicken *define each other*, they are co-relative. It is the ongoing process of living which has shaped our world in the back-and-forth between what we describe as external constraints from our perceptual perspective and the internally generated activity. The origins of this process are for ever lost, and our world is for all practical purpose stable (. . . except when it breaks down). But this apparent stability need not obscure a search for the mechanisms that brought them forth. It is this emphasis on co-determination (beyond chicken and egg) which marks the difference between the enactive viewpoint and any form of constructivism[41] or biological neo-Kantism.[42] This is important to keep in mind, since the more or less realist philosophy that pervades cognitive science will tend to assume that anybody who questions representations must *ipso facto* be in the antipodal position where the spectre of solipsism also lives.

### *Colour and Smell as Examples*

The preceding considerations are usually made in the realm of language and human communications, and it would seem that for the more immediate perceptual world they would not be relevant. But the whole point is that enaction applies at *all* levels. Thus examining perception at this light is important.

Consider the world of colours that we perceive every day. It is normally assumed that colour is an attribute of the wavelength of reflected light from objects that we pick it up and process it as relevant information. In fact, as has now been extensively documented, the perceived colour of an object is largely independent of the incoming wavelength.[43] Instead, there is a complex (and only partially understood) process of cooperative comparison between multiple neuronal ensembles in the brain, which specifies the colour of an object according to the global state it reaches: a perceptual chromatic space is specified.[44]

Now, clearly these mechanism are consistent with what we describe as illumination constraints (reflectance, object discontinuity, and so on) but they are not a *logical* consequence of them. The cooperative neuronal operations underlying our perception of colour, have resulted from the long biology evolution of the primate group. Their effects are so pervasive to our life that it is tempting to assume that colours, as we see them, is the way the world *is*. But this conclusion is tempered if we remember that other species have evolved different chromatic worlds by performing different cooperative neuronal operations from their sensory organs. For example, the pigeon's chromatic space is apparently tetrachromatic (requires four primary colours), in contrast to us trichromats (where only three primary colours suffice).[45] This is not a merely expansion in diversity within the same spectrum, but an entirely new dimension which brings forth a chromatic world as incommensurable to ours as ours is to a daltonic person. Colour here appears not as a correlate of world properties, but as regularities which are co-defined with a particular mode of being.

What can be said is that our chromatic world is *viable*: it is effective since we have continued our biological lineage. The vastly different histories of structural coupling of birds, insects, and primates have brought forth a world of relevance for each inseparable from their living. All that is required is that each path taken is viable, i.e. be an uninterrupted series of structural changes. The neuronal mechanisms underlying colour are not the solution to a 'problem' (picking up the correct chromatic properties of objects), but the arising together of colour perception *and* what one can then describe as chromatic attributes in the world inhabited.

Another perceptual dimension where these ideas can be seen at play is olfaction, not due to the comparative span provided by phylogeny, but due to novel electrophysiological techniques. Over many years of work, Freeman[46] has managed to insert an array of electrodes into the olfactory bulk of a rabbit so that a small portion of the global activity can be measured while the animal behaves freely. It was found that there is no clear pattern of global activity in the bulb unless the animal is exposed to one specific odor several times. Further, such emerging patterns seem to be created out of a background of incoherent activity into a coherent attractor. As in the case of colour, smell reveals itself not as a passive mapping of external traits, but as the creative dimensioning of meaning on the basis of history.[47]

In this light, then, the brain's operation is centrally concerned with the constant enactment of worlds through the history of viable lineages; an organ laying down worlds, rather than mirroring.

## *An Outline of the Doctrine*

The basic notion, then, is that cognitive capacities are inextricably linked to a history that is lived, much like a path that does not exist but is laid down in walking. Consequently, the view of cognition is not that of solving problems through representations, but as a creative bringing forth of a world where the only required condition is that it is *effective action*: it permits the continued integrity of the system involved,[48]

*Question # 1*: What is cognition?
*Answer*: Effective action: History of structural coupling which enacts (bring forth) a world.

*Question # 2*: How does it work?
*Answer*: Through a network of interconnected elements capable of structural changes undergoing an uninterrupted history.

*Question # 3*: How do I know when a cognitive system is functioning adequately?
*Answer*: When it becomes part of an existing on-going world of meaning (in ontogeny), or shapes a new one (in phylogeny).

It should be noted that two notions surface in these answers that are usually absent from considerations in CS. One is that, since representations no longer play a central role, intelligence has shifted from being the capacity to solve a problem to the capacity to *enter* into a shared world. The second is that what takes the place of task-oriented design is an *evolutionary* process. Bluntly stated, just as much as connectionism grew out of cognitivism inspired by a closer contact with the brain, the enactive orientation takes a further step in the same direction to encompass the temporality of living either in ontogeny and phylogeny.

### *Working without Representations*

Seeking alternatives to representation to study cognitive phenomena (and this is, admittedly, a vague umbrella, much as connectionism is) attracts a relatively small group of people in diverse fields. Further, as I shall argue in the next section, many of the tools of traditional connectionist perspective can be re-formulated in this context, so the divisory lines are much less sharp here than they were between the symbolic and connectionism orientations.

It is clear that an enactive strategy for AI is feasible only if one is prepared to relax the constraints of a specific problem solving performance. This is the spirit, for example, of the so-called classifier systems,[49] conceived to confront an undefined environment which it has to shape into significance. More generally, simulations of prolonged histories of coupling with various evolutionary strategies permit to discover trends wherein cognitive performances arise.[50] But these new perspectives for research are only at their earliest beginnings.

## 7. LINKING EMERGENCE AND ENACTION

The link between emergence and enaction depends on changing one's reading of what a distributed system can do. If one emphasizes how a historical process leads to emergent regularities without a fixed final constraint, one recovers the more open-ended biological condition. If one emphasizes, instead, how a given network will acquire a very specific capacity in a very definite domain (i.e. NetTalk), then representations are back in, and we have the more usual take on connectionist models. However, the first interpretation also entails a whole new different perspective on what cognition is, as outlined in the previous section.

Thus the road taken is strongly dependent on the degree of interest to stay closer to biological reality, and further away from a pragmatic-engineering considerations. Of course, defining a fixed domain within which a connectionist system can function is possible, but it obscures the deeper issues about origin so central to the enactive viewpoint.

Consider for example Smolesnky's Harmony theory. His viewpoint of sub-symbolic computation as a model for intuition seems eminently in line with an enactive perspective, which is why it can serve as the best case to consider here for contrast. However, even Harmony theory is evaluated in reference to an unviolated level of environmental reality: exogenous features matching given features of the world, and endogenous activity which acquire through experience a state of abstract meaning that "optimally encode environmental regularity". The hope is to find endogenous activity which corresponds to an "optimality characterization" of the surroundings.[51] The enactive perspective would require taking this kind of cognitive system into a situation where endogenous and exogenous are mutually definitory through a prolonged history requiring only a viable coupling, and eschewing any form of optimal fitness.[52]

Granted, from the standpoint of a pragmatically oriented AI, having as objective the production of a system that works in some domain in short delay, this orientation seems pointless. My argument is that cognitive properties emerged in living systems without such optimality considerations. They result from histories of viable compensations that create regularities, but it is far from obvious that they can be said to correspond to some unique referent.

Thus, there is a tension between the two parallel worlds of research, where the choice for or against the enactive critique is taken according to all the complexities of a conceptual shift, and the technological world, where the straight-jacket of immediate applicability sets the limits on how far it is able to extend itself. It seems to me that this tension will probably be resolved by a widening gap between the technological and the scientific components of CS.[53]

## 8. CONCLUSION: EMBODIED KNOWLEDGE

We started from the hard-core of CS and moved towards what might be considered its periphery, that is, the consideration of surrounding context, and effects of the biological and cultural history on cognition and action. Of course, those who hold on to representations as a key idea, see these concerns as only temporarily outside of the more precise realm of problem-solving orientation that seems more accessible; others go as far as to take the position that such 'fuzzy' and 'philosophical' aspects should not even enter into a proper cognitive science.

Some contrasts that provide these tensions may be stated in the following table.

| From: | Towards: |
|---|---|
| task-specific | creative |
| problem solving | problem definition |
| abstract, symbolic | history, body bound |
| universal | context sensitive |
| centralized | distributed |
| sequential, hierarchical | parallel |
| world pre-given | world brought forth |
| representation | effective action |
| implementation by design | implementation by evolutionary strategies |
| abstract | embodied |

As a visual summary of this presentation I have outlined the three main directions discussed here in a *polar map* of Figure 2. My view is

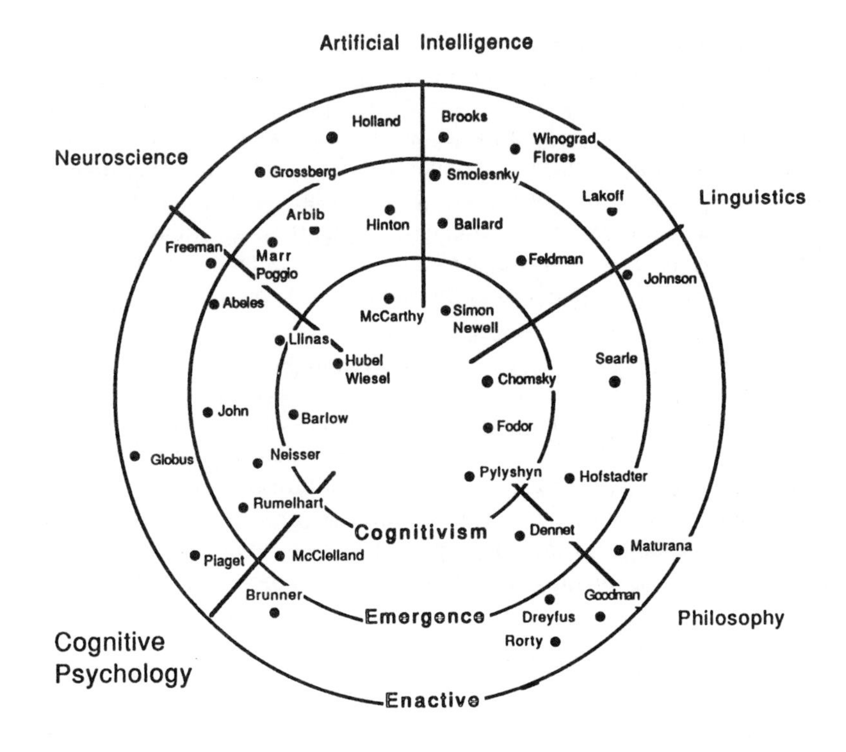


Fig. 2. A polar map of STC, having the cognitivist paradigm at the center and the alternative themes as fringe, both touching the double-edged body of connectionist ideas. Along the disciplinary radii, some names of workers in each representative area.

that these three successive waves to understand basic cognition and its origin relate to each other by successive *imbrication*, as Chinese boxes. In the centripetal direction, one goes from emergence to symbolic by bracketing the base from which symbols emerge, and working with symbols at face value. Or one can go from enaction to a standard connectionist view by assuming given regularities of the domain where the system operates (i.e. a fitness function in a domain). The centrifugal direction is a progressive bracketing of what seems stable and regular, to consider where such regularities could have come from, up to and including perceptual dimensions of our human world.

It is clear that each one of these approaches, as levels of descriptions, are useful in their own context. However, if our task is to understand the origin of perception and cognition as we find them in our actual lived history, I think that the correct level of explanation is the most inclusive outer rim of the map. Further, for an AI where machines are intelligent in the sense of growing into a common sense with human beings as animals can do, I can see no other route but to bring them up through a process of evolutionary transformations as suggested in the enactive perspective.[54] How fertile, how difficult, or how impossible this will prove to be, is anybody's guess.

My preferences have been quite explicitly laid out in the text. In particular, in this essay I have argue that, if the kingpin of cognition is its capacity for bringing forth meaning, then information is not pre-established as a given order, but it amounts to regularities that emerge from the cognitive activities themselves. It is this re-framing that has multiple ethical consequences, which should be evident by now. As pointed out in the last item of our table above, to the extent that we move from an abstract to a fully embodied view of knowledge, facts and values become *inseparable*. To know *is* to evaluate through our living, in a creative circularity.

*CREA, Ecole Polytechnique, Paris*

NOTES

[1] This Section owes much to our recent collective work on the neglected history of early cybernetics, self-organization, and cognition, published as *Cahiers du CREA* N°s 7—9. The only other useful source is S. Heims, *John von Neumann and Norbert Wiener*, MIT Press, 1980. The recent book by H. Gardner, *The Mind's New Science: A History of the Cognitive Revolution*, Basic Books, 1985, discusses this period only in a superficial way.

[2] The best sources here are the oft-cited Macy Conferences, published as *Cybernetics-Circular causal and feedback Mechanisms in Biological and Social Systems*, Josiah Macy Jr. Foundation, New York, 5 volumes.

[3] *Bulletin of Mathematical Biophysics*, **5**, 1943. Reprinted in W. McCulloch, *Embodiments of Mind*, MIT Press, 1965.

[4] For an interesting perspective about this historical/conceptual moment see also A. Hodges, *Alan Turing: The Enigma of Intelligence*, Touchstone, New York, 1984.

[5] See H. Gardner, *op. cit.*, Chapter 5 for this period.

[6] This designation is justified in J. Haugland (Ed.), *Mind Design*, MIT Press, 1981. Other designations used are: computationalism (Fodor) or symbolic processing. For this section I have profited much from D. Andler's article in *Cahier du CREA* N° 9.

[7] For more on this see J. Searle, *Intentionality*, Cambridge U. P., 1983.

[8] R. Shepard and J. Metzler, *Science* **171**, 701—703 (1971).

[9] S. Kosslyn, *Psychol. Rev.* **88**, 46—66, 1981.
[10] See *Beh. Brain Sci.* **2**, 535—581 (1979).
[11] This is the opening line of a popular textbook in neuroscience: "The brain is an unresting assembly of cells that continually receives information, elaborates and perceives it, and makes decision." S. Kuffler and J. Nichols, *From Neuron to Brain*, Sinauer Associates, Boston, 2nd ed., 1984, p. 3.
[12] D. Hubel and T. Wiesel, *J. Physiol.* **160**, 106 (1962). For a recent account of this work see Kuffler and Nichols, *op. cit.* Ch. 2—4.
[13] H. Barlow, *Perception* **1**, 371—392.
[14] F. Rosenblatt, *Principles of Neurodynamics: Perceptrons and the Theory of Brain Mechanisms*, Spartan Book, 1962.
[15] For more on the complex early origins of self-organization ideas see I. Stengers, *Cahier du CREA* N° 8, pp. 7—105.
[16] 'The logical geography of computational approaches', MIT Sloan Conference, 1984.
[17] For extensive discussion on this point of view see P. Dumouchel and J.-P. Dupuy (Eds.), *L'Auto-organisation: De la physique au politique*, Eds. du Seuil, Paris, 1983.
[18] See for example H. von Foerster (Ed.), *Principles of Self-Organization*, Pergamon Press, 1962.
[19] An accessible introduction to the modern theory of dynamical systems is: R. Abraham and C. Shaw, *Dynamics: The Geometry of Behavior*, Aerial Press, Santa Cruz, 3 vols., 1985.
[20] G. Horn and R. Hill, *Nature* **221**, 185—187 (1974).
[21] M. Fishman and C. Michael, *Vision Res.*, **13**, 1415 (1973) and F. Morell, *Nature* **238**, 44—46 (1972).
[22] J. Allman, F. Miezen and E. McGuiness, *Ann. Rev. Neuroscien.* **8**, 407—430 (1985).
[23] F. Varela and W. Singer, *Exp. Brain Res.* **66**, 10—20 (1987).
[24] An interesting collection of examples is: G. Palm and A. Aersten (Eds.), *Brain Theory*, Springer Verlag, 1986.
[25] The name is proposed in: J. Feldman and D. Ballard, 'Connectionist models and their properties', *Cognitive Science* **6**, 205—254 (1982). For extensive discussion of current work in this direction see: D. Rumelhart and J. McClelland (Eds.), *Parallel Distributed Processing: Studies on the Microstructure of Cognition*, MIT Press, 1986, 2 vols.
[26] The main idea is due to J. Hopfield, *Proc. Natl. Acad. Sci.* (U.S.A.), **79**, 2554—2556 (1982).
[27] There are many variants associated to these ideas. See in particular: G. Hinton, T. Sejnowsky, and D. Ackley, *Cognitive Science* **9**, 147—163 (1984), and G. Toulousse, S. Dehaene, and J. Changeaux, *Proc. Natl. Acad. Sci.* (U.S.A.), **83**, 1695—1698 (1986).
[28] The idea is due to D. Rumelhart, G. Hinton, and R. Williams, in: Rumelhart and McClelland, *op. cit.*, Ch. 8.
[29] T. Sejnowski and C. Rosenbaum, 'NetTalk: A parallel network that learns to read aloud', TR JHU/EECS-86/01, John Hopkins Univ.
[30] For the distinction between symbolic and emergent description and explanation in biological systems see F. Varela, *Principles of Biological Autonomy*, North Holland, New York, 1979, Ch. 7, and more recently S. Oyama, *The Ontogeny of Information*, Cambridge U. Press, 1985.

[31] See D. Hillis, 'Intelligence as emergent behavior', *Daedalus*, Winter 1989, and P. Smolesnky, 'On the proper treatment of connectionism', *Beh. Brain Sci.* **11**: 1, 1989. In a very different vein J. Feldman, ' Neural representation of conceptual knowledge', U. Rochester TR189 (1986) proposes a middle ground between 'punctuate' and distributed systems.
[32] P. Smolesnky in: Rumelhart and McClelland, *op. cit.*, Ch. 6.
[33] This is extensively argued by two noted spokesmen of cognitivism: J. Fodor and S. Pylyshyn, 'Connectionism and cognitive architecture: A critical review', *Cognition*, 1989. For the opposite philosophical position in *favor* of connectionism see: H. Dreyfus, 'Making a mind vs. modeling the brain: AI again at the cross-riads'. *Daedalus*, Winter, 1989.
[34] Most influential in this respect is the work of H. G. Gadamer, *Truth and Method*, Seabury Press, 1975. For a clear introduction to hermeneutics see Palmer, *Hermeneutics*, Northwestern Univ. Press, 1979. The formulation of this section owes a great deal to the influence of F. Flores, see: T. Winnograd and F. Flores *Understanding Computers and Cognition: A New Foundation for Design*, Ablex, New Jersey, 1986.
[35] The name is far from being an established one. I suggest it here for pedagogical reasons, until a better one is proposed.
[36] H. Dreyfus and S. Dreyfus, *Mind over Machine*, Free Press/Macmillan, New York, 1986.
[37] For this explicit way of constructing biologically inspired networks see T. Poggio, V. Torre and C. Koch, *Nature* **317**, 314—319 (1986).
[38] For an interesting sample of discussion in AI about these themes see the multiple review of Winnograd and Flores's book, in *Artif. Intell.* (1987).
[39] The main reference points we have in mind here are (in their English versions): M. Heidegger, *Being and Time*, Harper and Row, 1977; M. Merleau-Ponty, *The Phenomenology of Perception*, Routledge and Kegan Paul, 1962; Michel Foucault, *Discipline and Punish: The birth of the prison*, Vintage/Random House, 1979.
[40] This is discussed in my contribution to the previous Stanford Symposium, 'Living ways of sense making: A middle way approach to neuroscience', in P. Livingston (Ed.), *Order and Disorder*, Anma Libris, Stanford, 1984.
[41] See for instance P. Watzlawick (Ed.), *The Invented Reality: Essays on Constructivism*, Norton, New York, 1985.
[42] Most clearly seen in the Vienna school of Konrad Lorenz, as expressed, for example, in *Behind the Mirror*, Harper and Row, 1979.
[43] See for instance E. Land, *Proc. Natl. Acad. Sci.* (U.S.A.) **80**, 5163—5169 (1983).
[44] P. Gouras and E. Zenner, *Progr. Sensory Physiol.* **1**, 139—179 (1981).
[45] F. Varela *et al., Arch. Biol. Med. Exp*, **16**, 291—303 (1983); E. Thompson, A. Palacios, F. Varela, *Beh. Brain Sci.* (in press), 1992.
[46] W. Freeman, *Mass Action in the Nervous System*, Academic Press, 1975.
[47] W. Freeman and C. Skarda, *Brain Res. Reviews*, **10**, 145—175 (1985). Significantly, a section of this article is entitled: 'A retraction on "representation"' (p. 169).
[48] This biologically inspired re-interpretation of cognition was presented in H. Maturana and F. Varela, *Autopoiesis and Cognition: The realization of the Living*, D. Reidel, Boston, 1980, and F. Varela, *Principles of Biological Autonomy, op. cit.* For an introductory exposition to this point of view and more recent developments see H. Maturana and F. Varela, *The Tree of Knowledge: the Biological Roots of Human Understanding*,

New Science Library, Boston 1987. The links with language and AI are discussed in Winnograd and Flores, *op. cit.*

[49] See J. H. Holland, 'Escaping brittleness', in: *Machine Intelligence*, Vol. 2 (1986).

[50] An interesting recent collections of diverse papers in this direction can be found in: *Evolution, Games and Learning: Models for Adaptation in Machines and Nature, Physica 22D* (1986). Surely, many of the contributors would not agree with our readings of their work. For an explicit example see: F. Varela, 'Structural coupling and the origin of meaning in a simple cellular automata', in E. Secarz, (Ed.), *The Semiotics of Cellular Communications*, Springer-Verlag, New York, 1987.

[51] P. Smolesnky, *op. cit.*, p. 260.

[52] It is worth noting that similar arguments can be applied to evolutionary thinking today. For the parallels between cognitive representationism and evolutionary adaptationism, see F. Varela, in: P. Livingston (Ed.), *op. cit.* and the Introduction in this volume.

[53] See also the remarks by Roger Schank in *AI Magazine*, pp. 122—135 (Fall 1985).

[54] This is the trend within the new field of 'Artificial Life'; see e.g. Ch. Langton (Ed.), *Artificial Life*, Addison-Wesley, New Jersey, 1990.

CHRISTINE A. SKARDA

# PERCEPTION, CONNECTIONISM, AND COGNITIVE SCIENCE

## I. INTRODUCTION

Recent findings in neurophysiology and cognitive science point to the same conclusion: cognition can be explained without appeal to the representations and rules of earlier cognitivist explanations. Yet if this is true, we want to know what form the alternative explanation will take and what processes are responsible for cognitive phenomena like perception. In this paper I discuss three issues: (1) the correct characterization of the alternative to cognitivism; (2) the resulting view of perception based on the alternative; and (3) the implications of this alternative explanatory framework for cognitive science.

## II. CONNECTIONIST OPTIONS

Varela offers two options to traditional 'cognitivism' (the view that cognition can be explained in terms of formal symbol manipulation). One of these options he terms 'emergence', the other 'enaction'. Emergent systems are equated with current connectionist approaches, while enactive systems are viewed as a non-connectionist, second option to cognitivism. I would like to reformulate this: there is an important distinction to be made here, but it is not the one Varela makes in his paper.

In a recent paper (Freeman and Skarda, 1988), Walter Freeman and I suggest that it is important to distinguish two camps of present-day connectionist models, one typified by so-called PDP systems (Hinton, 1985; Rummelhart *et al.*, 1986), the other characterized by self-organizing dynamical systems (Amari, 1983; Freeman, 1975; Grossberg, 1981; Hopfield, 1982; Kohonen, 1984). Varela conflates these two classes of connectionist models in his paper when he describes connectionist systems as *distributed systems* that are also *self-organized*.

Distributed systems are not *eo ipso* self-organizing systems. Systems that fall within the PDP class of connectionist models use *globally*



*Francisco J. Varela and Jean-Pierre Dupuy (eds), Understanding Origins,* 265—271.

*distributed dynamics*, but there is a sense in which this class of systems still uses internal representations in the production of behavior. Systems like these, that rely on feed-forward connectivity and back propagation for error correction, have their 'goals' externally imposed. As Varela points out, they require a 'teacher' or set of correct answers to be introduced by the system's operator. These answers are paradigmatic patterns with reference to which the output of the system is corrected via error correction. The teacher may not be contained in a program, but it functions in the same way as an internal representation does in conventional computers.

Some connectionist systems, however, are self-organized systems. Self-organizing dynamic systems, because of dense local feedback connections, do not require or use teachers. No matching or comparison takes place such as by correlation or completion, and no archetypal set patterns are placed by an external operator into the system as its goals. Self-organized systems do not fall prey to the criticisms Varela levels at 'connectionist' systems.

It is misleading to identify, as Varela does, connectionism with self-organizing, *emergentist* systems, and to say that *all* connectionist systems are still wedded to the representations of traditional cognitivism. Some connectionist models are self-organizing, but others are not. All connectionist systems use distributed, highly parallel processing, but that is not the same thing as being self-organizing. PDP systems are susceptible to Varela's attack on representations, self-organized systems are not. I believe that Varela's distinction between emergent and enactive systems is ultimately intended to capture the same fundamental distinction, but it is mistaken to equate emergent systems with connectionism as a whole and set all connectionist systems against the enactive approach. This dichotomy is a false one.

I believe that this point of clarification is important for two reasons. First, if Varela would draw the distinction as I have rather than as he has, his own position would be strengthened by being corroborated by an important class of connectionist models. Second, historically calls for an alternative to cognitivism have been beset by vagueness concerning what form the alternative would take. In order to adopt a nonsymbolic, nonrepresentational approach to cognition we need more than the phenomenologists of the continental tradition have produced, more than discussions of why knowledge is a matter of 'being in a world', more than Varela's claim that cognition is an "on-going interpretation which cannot be adequately captured as a set of rules and assumptions since it is a matter of action and history, an understanding picked up by imitation and by becoming a member of an understanding which is already there". As cognitive scientists we want a model of how this self-organized interaction works. Self-organizing connectionist systems are a step in the direction of defining a nonrepresentational alternative in cognitive science, and as such they are of crucial importance.

## III. REDEFINING PERCEPTION

The recognition of self-organizing systems is important because it forces a redefinition of perception along the lines sketched by Varela under the term 'enaction'. Data gathered in Freeman's laboratory has led to similar conclusions and serves as a concrete example in the present context.

Both symbol-based and distributed PDP models of perception view perception as a process initiated by the causal impact of an object on the system that leads to the formation of a more or less adequate internal representative of that object and its features. Perception on this model is a reaction to something that is initiated at the receptor level, it is pick-up, detection, representation of some object or state of affairs. Varela neatly summarizes this position in his paper.

Investigation of sensory processing in the olfactory bulb leads to a very different picture of perception (Freeman and Skarda, 1985). Neural dynamics in the bulb are self-organizing. Evidence indicates that when an organism is trained to respond to a particular odor a self-organized process in the bulb produces a spatially coherent state of patterned activity that can be modelled mathematically as a limit cycle attractor. With each inhalation, after learning and in the presence of this odor, this more ordered state repeatedly emerges from the background state which itself is self-organized. A separate spatial pattern of periodic behavior forms for each odor given under reinforcement. When the reinforcement contingency is changed in respect to any one odor, or if a new odor is added to the repertoire under reinforcement, all the spatial patterns undergo small changes during the process of learning. These changes do not occur in the olfactory bulb if there is no reinforcement or if the newly learned CS is not olfactory but visual or auditory.

With respect to perception several features of the neural dynamics are worthy of note. (1) Only when the odorant is reinforced leading to a behavioral change, i.e. only when the stimulus input has some behavioral significance for the organism such that it acts on the stimulus, do odor-specific activity patterns form in the olfactory bulb. Presentation of odorants to the receptors in unmotivated subjects does not lead to any observable changes in the system. (2) Odor-specific activity patterns are dependent on the behavioral response: changing the reinforcement contingency changes the patterned activity previously recorded. (3) The self-organized, internally generated patterned activity is context dependent: introducing new reinforced ordorants to the animals' repertoire leads to changes in the patterns of all previously learned odorants.

These findings have important implications for how we view perception. First, perception does not begin with causal impact on receptors; it begins within the organism with internally generated (self-organized) neural activity that, by re-afference, lays the ground for processing of future receptor input. In the absence of such activity, receptor stimulation does not lead to any observable changes in neural dynamics in the brain. It is the brain itself that creates that conditions for perception by generating activity patterns that determine what receptor activity will count for it. Perception is interaction initiated by the organism, not reaction caused by the object at the receptor level. Thus, the story of perception cannot be told simply in terms of feed-forward causation in which the object initiates neural changes leading to an internal perceptual state. What is missing in the reflex-based model is recognition of the role played by self-organized neural processes and by dense feedback among subsystems in the brain that allow the organism to initiate interaction with its environment.

Second, the fact that odor-specific activity patterns change whenever new odors are added to the repertoire or when reinforcement contingencies (behavioral responses) are altered, indicates that perception is not internal representation of an object. The self-organized neural activity we record reflects a process of reliable interaction in a context. These patterns reflect not just the presence of an odorant, or the response, but both in interaction along with a context of other significant odorants in which this behavior is embedded.

These findings provide neurophysiological support for a nonrepresentational, cognitive alternative for cognitive science. Varela's critique of problem solving and representation, and his emphasis on what he terms 'enaction', I take to be another way of getting at a view of perception that Freeman and I have developed on the basis of olfactory processing.

## IV. IMPLICATIONS FOR COGNITIVE SCIENCE

The self-organized property of brain dynamics also has important implications for cognitive science because self-organized systems require an explanatory framework alien to that used traditionally in science (Skarda, 1986). Explanations, including those in cognitive science, have been attempts to understand system properties in terms of the properties of the input to the system and of the parts that constitute the system. Yet, explanations of self-organizing phenomena can only be given in terms of qualitative forms of behavior of the system as a whole. These system properties resist analysis in terms of the properties of the parts that comprise the system or in terms of properties of the input to the system. In explaining such phenomena there is relative independence from the nature and properties of the substrate; hence microreduction, the aim of traditional explanations, does not work (Garfinkel, 1981). Cognitive science must take this into account.

The requirement for a new explanatory model to deal with brain dynamics underlying perception and behavior is further underscored by the role played by chaotic neural activity. The term 'chaos' refers to dynamic activity that appears random, but is not. Such activity exists in many forms and degrees, has precisely definable characteristics and relatively few degrees of freedom, and can be reliably simulated, generated, or reproduced if initial conditions are identical and known. Chaotic activity has been identified in more than one brain area (Babloyantz and Destexhe, 1986; Nicolas and Tsuda, 1985; Freeman and Viana Di Prisco, 1986; Garfinkel, 1983), and we have postulated that it may provide the basis for the flexibility and adaptive coping that make possible successful interaction with an unpredictable environment (Skarda and Freeman, 1987)

The observation that brains employ chaos to produce behavior is important in the present context because it is known that chaotic phenomena preclude all long-term predictions. It may seem paradoxical to make this claim about a deterministic phenomenon, but in systems that exhibit chaotic behavior small uncertainties are amplified by the nonlinear interactions of a few elements. The upshot is that behavior that was predictable in the short run become intrinsically unpredictable in the long term (Crutchfield *et al.*, 1987). As a result, physiologists cannot make strict casual inferences from the level of individual neurons to that of neural mass actions, nor from the level of receptor activity to internal dynamics. Thus chaotic, self-organized systems challenge the traditional reductionist paradigm of explanation that lies at the heart of cognitive science. Qualitative descriptions of global system dynamics replace the reductive explanations of the past. Moreover, these phenomena make long-term predictions intrinsically impossible, they cut the causal connection between past and future. Taken together, these facts imply a change in the nature of explanations in cognitive science and a new direction for future research.

*CREA, Ecole Polytechnique, Paris*

## REFERENCES

Amari, S. : 1983, 'Field theory of self-organizing neural nets', *IEEE Transactions on Systems, Man, and Cybernetics* SMC-13 No. 5, 741—749.

Babloyantz, A. and Destexhe, A. : 1986, 'Low-dimensional chaos in an instance of epilepsy', *Proc. Nat. Acad. Sci. U.S.A.* **83**, 3513—3517.

Crutchfield, J., Farmer, J., Packard, N. and Shaw, R. : 1987, 'Chaos', *Scientific American* **255**, 46—57.

Freeman, W. : 1975, *Mass Action in the Nervous System*. Academic Press.

Freeman, W. and Skarda, C. : 1985, 'Spatial EEG patterns, nonlinear dynamics and perception: The neo-Sherringtonian view', *Brain Research Reviews* **10**: 147—175.

Freeman, W. and Skarda, C. : 1988, 'Mind/brain science: neuroscience on philosophy of mind', in: LePore, E. and van Gulick, R. (Eds.) *Festschrift for John R. Searle*. Oxford: Blackwell.

Freeman, W. and Viana Di Prisco, G.: 1986, 'EEG spatial pattern differences with discriminated odors manifest chaotic and limit cycle attractors in olfactory bulb of rabbits', in: Palm, G. (Ed.) *Brain Theory*. New York: Springer Verlag.

Garfinkel, A. : 1981, *Forms of Explanation*. Yale University Press.

Garfinkel, A. : 1983, 'A mathematics for physiology', *American Journal of Physiology* **245** (Regulatory, Integrative and Comparative Physiology) (14), R455—466.

Grossberg, S. : 1981, 'Adaptive resonance in development, perception, and cognition', in: Grossberg, S. (Ed.) *Mathematical Psychology and Psychophysiology*. American Mathematical Society.

Hinton, G. : 1985, 'Learning in parallel networks', *Byte* **10**, 265.

Hopfield, J. : 1982, 'Neural networks and physical systems with emergent collective computational abilities', *Proc. Nat. Acad. Sci. U.S.A.* **79**: 2554.

Kohonen, T. : 1984, *Self-Organization and Associative Memory*. New York: Springer Verlag.

Nicholas, J. and Tsuda, I. : 1985, 'Chaotic dynamics of information processing. The "magic number seven plus-minus two" revisited', *Bulletin of Mathematical Biology* **47**, 343–365.

Rumelhart, D., McClelland, J. and PDP Research Group: 1986, *Parallel Distributed Processing: Explorations in the Microstructures of Cognition. Vol. I: Foundations.* MIT Press/Bradford.

Skarda, C. : 1986, 'Explaining behavior: Bringing the brain back in', *Inquiry* **29**, 187–202.

Skarda, C. and Freeman, W. : 1987, 'How brains make chaos in order to make sense of the world', *Behavioral and Brain Sciences* **10**, 161–173.

UMBERTO ECO

# THE ORIGINAL AND THE COPY*

It seems that in terms of natural language everybody knows what a fake, a forgery or false document is. At most, one admits that it is frequently difficult to recognize a forgery as such, but one relies on experts, that is, on those who are able to recognize forgeries simply because they know how to tell the difference between a fake and its original.

As a matter of fact, the definitions of such terms as fake, forgery, pseudepigrapha, falsification, facsimile, counterfeiting, spurious, pseudo, apocryphal and others, are rather controversial. It is reasonable to suspect that many difficulties in defining these terms are due to the difficulty in defining the very notion of 'original' or of 'real object'.

## 1. PRELIMINARY DEFINITIONS

### 1.1. *Current Definitions*

Here follow some definitions from the *Webster's New Universal Unabridged Dictionary.*

*Forgery*: "the act of forging, fabricating or producing falsely; especially, the crime of fraudulently making, counterfeiting, or altering any writing, record, instrument, register, note and the like to deceive, mislead or defraud; as the forgery of a document or of a signature".

*Fake* (v. t.): "to make (something) seem real, satisfactory, etc., by any sort of deception; to practice deception by simulating or tampering with (something); counterfeit (Colloq.)".

*Fake* (n.): "anything or anyone not genuine".

*Facsimile*: "any copy of likeness".

*Spurious*: "illegitimate, bastard .... False; counterfeit; not genuine ... in botany: false, like in appearance but unlike in structure or function



*Francisco J. Varela and Jean-Pierre Dupuy (eds), Understanding Origins,* 273—303.

(*spurious primary* or *quill*: the outer primary quills when rudimentary or very short, as in certain singing birds). Syn.: counterfeit, fictitious, apocryphal, false, adulterate, bastard".

*Pseudo*: "fictitious, pretended, sham (as in pseudonym); counterfeit, spurious, as in pseudepigrapha; closely or deceptively similar to (a specified thing) as in pseudomorph; not corresponding to the reality, illusory . . . "

*Apocryphal*: "various writings falsely attributed . . . of doubtful authorship or authenticity . . . spurious".

A short inspection in other linguistic territories does not offer any more satisfactory help. Moreover, the term apocryphal (etymologically: secret, occult) designated at the beginning of the Christian era non-canonical books kept out from the New Testament, while Pseudepigrapha were writings falsely attributed to Biblical characters. For Protestants, the Apocrypha are in general fourteen books of the Septuagint regarded as non-canonical. Since, however, Catholics accept in the Roman canon eleven of these fourteen books, calling them Deuterocanonical, and call apocrypha the remaining three, then for Protestants the Catholic deuterocanonical books are usually called apocrypha and the Catholic apocrypha are called pseudoepigrapha.[1]

It is evident that all these definitions can work only once one has duly interpreted such terms as false, deceiving, misleading, fictitious, illusory, non-corresponding to reality, pretended, fraudulent, adulterated, as well as genuine, real, satisfactory, similar, and so on. Each of these terms is obviously crucial for a semiotic theory and all together they depend on a 'satisfactory' semiotic definition of Truth and Falsity.

It seems however rather difficult to look for a definition of Truth and Falsity in order to reach (after some thousands of pages of a complete revisitation of the whole course of Western and Eastern philosophy) a 'satisfactory' account of fakes. The only solution is thus to try a provisional and commonsensical definition of /forgery/and/fake/ — in order to cast in doubt some of our definitions of Truth and Falsity.

### 1.2. *Primitives*

In order to outline a provisional definition of forgery and fake we must take as primitive such concepts as similarity, resemblance and iconism. These concepts are discussed and defined in Eco (1976, 3.5 and 3.6).

Another concept we shall take as a primitive is the one of *identity* (as a criterion of identity of things, not of terms, concepts or names). Let us assume as a starting point Leibniz's law of the *identity of indiscernibles*: if, given to objects *A* and *B*, everything that is true of *A* is also true of *B*, and *vice versa*, and if there is no discernible difference between *A* and *B*, that is, innumerable 'properties' can be predicated of the same object, let us assume, that, rather than in the predication of those substantial properties advocated by Aristotle (*Met.* v, 9, 1018a: "things whose matter is formally or numerically one, and things whose substance is one, are said to be the same"), we are interested in the predication of a crucial 'accidental' property: two supposedly different things are discovered to be the same if they succeed in occupying at the same moment the same portion of space (for space-temporal identity see Barbieri, 1987, 2; for trans-world-identity see Hintikka, 1969, Rescher, 1973, Eco, 1979, 8.6.3).

Such a test is however insufficient for forgeries because we normally speak of forgeries when something present is displayed as if it were the original, while the original (if any) is elsewhere. One is thus unable to prove that there are two different objects occupying at the same time two different spaces. If by chance one is in the position of perceiving at the same time two different even though similar objects, then one is certainly able to detect that each of them is identical with itself and that they are not indiscernibly identical, but no criterion of identity can help to identify the original one.

Thus, even if we start from the above primitive concepts, we shall be obliged to outline additional criteria for distinguishing authentic from fake objects.

The many problems elicited by such an attempt will arouse some embarrassing suspicions about several current philosophical and semiotical notions, for example, originality and authenticity, as well as about the very concepts of identity and difference.

## 2. REPLICABILITY OF OBJECTS

It appears from the above current definitions that fakes, forgeries and the like concern cases in which either (i) there is a physical object that, because of its similarity with some other object, can be mistaken for it, or (ii) a given object is falsely attributed to an author who is said to have made — or supposed to have been able to make — similar objects.

It remains however unprejudiced whether these mistakes are caused by someone who had the intention of deceiving, or are accidental and fortuitous (see Section 3). In this sense a forgery is not an instance of lie through objects. At most, when a fake is presented as if it were the original with the explicit intention of deceiving (and not by mistake), there is a lie uttered about that object.

A semiotics of the lie is undoubtedly of paramount importance (see Eco, 1976, 0.1.3) but when dealing with fakes and forgeries we are not directly concerned with lies. We are first of all concerned with the possibility of mistaking one object for another because they share some common features.

In our everyday experience the most common case of mistakes due to similarity is the one in which we hardly distinguish between two tokens of the same type, as when in the course of a party we have put our glass down somewhere, next to another one, and are later unable to identify it.

### 2.1. *Doubles*

Let us define as a *double* a physical *token* which possesses all the characteristics of another physical *token*, at least from a practical point of view, in so far as both possess all the essential attributes prescribed by an abstract *type*. In this sense two chairs of the same model or two pieces of typing paper are each the double of the other and the complete homology between the two objects is established by reference to their type.

A double is not identical (in the sense of indiscernibility) with its twin, that is, two objects of the same type are physically distinct from one another: nevertheless, they are considered to be *interchangeable*.

Two objects are doubles of one another when for two objects $O_a$ and $O_b$ their material support displays the same physical characteristics (in the sense of the arrangement of molecules) and their shape is the same (in the mathematical sense of 'congruence'). The features to be recognized as similar are determined by the type. But who is *to judge* the criteria for similarity or sameness? The problem of doubles seems to be an ontological one, but it is rather a pragmatic one. It is the user who decides the 'description' under which, according to a given practical purpose, certain characteristics are to be taken into account in determining whether two objects are 'objectively' similar and consequently interchangeable. One need only consider the case of industrially produced and commercially available fakes: the reproduction does not possess all the features of the original (the material used may be of lower quality, the form may not be precisely the same), but the buyer displays a certain flexibility in the evaluation of the essential characteristics of the original and considers — whether from thriftiness, snobbery or indifference — the copy as adequate for his needs, either for consumption or for display. The recognition of doubles is a pragmatic problem, because it depends on cultural assumptions.

### 2.2. *Pseudo Doubles*

There are cases in which a single token of a type acquires for some users a particular value, for one or more of the following reasons:

(i) *Temporal priority.* For a museum or for a fanatic collector the first token of the Model T produced by Ford is more important than the second one. The coveted token is not different from the others and its priority can only be proved on the grounds of external evidence. In certain cases there is a formal difference due to imperceptible (and otherwise irrelevant) features: for example when only the first or a few early copies of a famous incunabulum are affected by a curious typographical imperfection that, since it was later corrected, proves the temporal priority of this or these copies.

(ii) *Legal priority.* Consider the case of two 100 dollar bills with the same serial number. Clearly, one of them is a forgery. Suppose that one is witnessing a case of 'perfect' forgery (no detectable differences in printing, paper, colors and watermark). It should be ascertained which one was produced at a given precise moment by an authorized maker. Suppose now that both were produced at the same moment in the same place by the Director of the Mint, one on behalf of the Government and the other for private and fraudulent purposes. Paradoxically, it would be sufficient to destroy either bill, and to appoint as legally prior the surviving one.

(iii) *Evident association.* For rare book collectors, an 'association copy' is one which bears the signature of the author or any owner's mark of a famous person (obviously these evidences can be forged in their turn). Normally two bank notes of the same denomination are considered interchangeable by normal people, but if a given bank note marked with the serial number *x* was stolen in the course of a bank robbery, this and only this one becomes significant for a detective who wants to prove someone guilty.

(iv) *Alleged association.* A token becomes famous because of its supposed (but not physically evident) connection with a famous person. A goblet which is in outward appearance interchangeable with countless others, but was the one used by Jesus Christ at the Last Supper, becomes the Holy Grail, the unique target of an unending Quest. If the Grail is merely legendary, the various beds in which Napoleon slept for a single night are real and are actually displayed is many places.

(v) *Pseudo association.* This is a case in which a double looks like a pseudo double. A great number of tokens of the same industrial type (be they bags, shirts, ties, watches and so on) are coveted because they bear the emblem of a famous producer. Each token is naturally interchangeable with any other of the same kind. It can happen however that another minor company makes perfect tokens of the same type, with no detectable differences in form and matter and wish a forged emblem reproducing the original one. Any difference should only concern lawyers (it is a typical case of merely legal priority) but many customers, when realizing that they have bought the 'wrong' token, are as severely disappointed as if they had obtained a serial object instead of a unique one.

### 2.3. *Unique Objects with Irreproducible Features*

There are objects so complex in material and form that no attempt to reproduce them can duplicate all the characteristics acknowledged as essential. This is the case with an oil painting done with particular colors on a particular canvas, so that the shades, the structure of the canvas and the brush strokes, all essential in the appreciation of the painting as a work of art, can never be completely reproduced. In such cases a unique object becomes *its own type* (see Section 5, and the difference between *autographic* and *allographic* arts). The modern notion of a work of art as irreproducible and unique assigns a special status both to the origin of the work and to its formal and material complexity, which together comprise the concept of *authorial authenticity.*

Frequently, in the practice of collectors, the temporal priority becomes more important than the presence of irreproducible features. Thus in statuary, where it is sometime possible to cast a copy which possesses all the features of the original, temporal priority plays a crucial role, even though the original may have lost some of its features (for instance, the nose is broken) while the copy is exactly as the original originally was. In such cases one says that artistic fetishism prevails over aesthetic taste (see 4.1.4 and the difference between the Parthenon of Athen and the one of Nashville).

## 3. FORGERY AND FALSE IDENTIFICATION

From a legal point of view, even doubles can be forged. But forgeries become semiotically, aesthetically, philosophically and socially relevant when they concern irreproducible objects and pseudo-doubles, in so far as both possess at least one external or internal 'unique' property. By definition, a unique object can have no double. Consequently any copy of it either is honestly labeled as a facsimile or is erroneously believed to be indiscernibly identical with its model. Thus a more restricted definition of forgery could be expressed as: any object which is produced — or, once produced, used or displayed — with the intention of making someone believe that it is indiscernibly identical with another unique object.

In order to speak of forgery it is necessary but not sufficient that a given object look absolutely similar to another (unique) one. It could happen that a natural force shapes a stone so as to transform it into a perfect copy or an indistinguishable facsimile of Michelangelo's *Moses*, but nobody, in terms of natural language, would call it a forgery. To recognize it as such it is indispensable that someone asserts that this stone is the 'real' statue.

Thus the *necessary* conditions for a forgery are that, given the actual or supposed existence of an object $O_a$, made by $A$ (be it a human author or whatever) under specific historical circumstances $t_1$, there is a different object $O_b$, made by $B$ (be it a human author or whatever) under circumstances $t_2$, which under a certain description displays strong similarities with $O_a$ (or with a traditional image of $O_a$). The *sufficient* condition for a forgery is that it be claimed by some Claimant that $O_b$ is indiscernibly identical with $O_a$.

The current notion of forgery generally implies a specific intention on the part of the forger, i.e. it presupposes *dolus malus*. However the question whether *B*, the author of $O_b$, was guilty of *dolus malus* is irrelevant (even when *B* is a human author). *B* knows that $O_b$ is not identical with $O_a$, and s/he may have produced it with no intention to deceive, either for practice or as a joke, or even by chance. Rather, we are concerned with any Claimant who claims that $O_a$ is identical with $O_b$ or can be substituted for it — though of course the Claimant may coincide with *B*.

However, not even the Claimant's *dolus malus* is indispensable, since s/he may honestly believe in the identity s/he asserts.

Thus a forgery is always such only for an external observer — the Judge — who, knowing that $O_a$ and $O_b$ are two different objects, understands that the Claimant, whether viciously or in good faith, has made a false identification.

According to some scholars, the *Constitutum Constantini* (perhaps the most famous forgery in Western history) was not initially produced as a false charter but as a rhetorical exercise. As in the course of the following centuries it was mixed with other types of document, it was step by step taken seriously by naive or fraudulent supporters of the Roman Church (De Leo, 1974). While it was not a forgery for the former, it was such for the latter, as it was for those who later started challenging its authenticity.

Something is not a fake because of its internal properties but by virtue of a *claim of identity*. Thus forgeries are first of all a *pragmatic* problem.

Naturally the Judge, the Claimant and both Authors are abstract roles, or *actants*, and it can happen that the same individual can play all of them at different times. For example, the painter *X* produces as Author *A* an Object *A*, then copies his first work by producing a second Object *B*, and claims that Object *B* is Object *A*. Later *X* confesses his fraud and, acting as the Judge of the forgery, demonstrates that Object *A* was the original painting.

## 4. PRAGMATICS OF FALSE IDENTIFICATION

We should exclude from a typology of false identification the following cases:

(i) *Pseudonymity*. To use a pen name means to lie (verbally) about the author of a given work, not to suggest identity between two works. Pseudonymity is different from pseudepigraphical identification (see

4.3), where the Claimant ascribes a given work $O_b$ to a well known or legendary author.

(ii) *Plagiarism.* In producing an $O_b$ which fully or partially copies an $O_a$, $B$ tries to conceal the similarity between the two objects, and does not try to prove their identity. When a Claimant says that the two objects are identical, s/he acts as a Judge and says so not in order to deceive anybody but rather in order to uncover $B$'s manoeuvre. When $B$ makes his/her dependency on $A$'s work evident, there is no plagiarism but rather parody, pastiche, homage, intertextual citation — none of these being an instance of forgery. A variation of these examples of pseudo-plagiarism are the works made *à la manière de* (see 4.3).

(iii) *Aberrant decoding* (see Eco, 1976: 142): when a text $O$ was written according to a code $C_1$ and it is interpreted according to a code $C_2$. A typical example of aberrant decoding is the oracular reading of Virgil during the Middle Ages or the erroneous interpretation of Egyptian hieroglyphs by Athanasius Kircher. Here one is not concerned with the identification between two objects but rather with different interpretation of a single one.

(iv) *Historical forgery.* In diplomatics there is a distinction between *historical forgery* and *diplomatic forgery.* While the latter is a case of forgery (see 4.3.1), the former is a mere case of lie. There is historical forgery when in an original document, produced by an author who is entitled to do so, something is asserted which is not the case. A historical forgery is not dissimilar from a false piece of news published by a newspaper. In this case (see Section 5) the phenomenon affects the content but not the expression of the sign function.[2]

Let us now consider three main categories of False Identification, namely Downright Forgery, Moderate Forgery and Forgery Ex-Nihilo.

### 4.1. *Downright Forgery*

We must presuppose that $O_a$ exists somewhere, that it is the unique original object, and that $O_a$ is not the same as $O_b$. Certainly such assumptions sound rather committing from an ontological point of view, but in this paragraph we are dealing with what the Claimant knows and we must take such knowledge for granted. Only in Section 6 shall we escape such an ontological commitment by discussing the criteria of identification to be used by the Judge.

Additional requirements are:

(i) the Claimant knows that $O_a$ exists and knows — or presumes to know on the grounds of even a vague description — what $O_a$ looks like (if a Claimant comes across *Guernica* and believes it is the *Mona Lisa* — which s/he has never seen nor has any clear idea about — then one is witnessing a single case of misnaming).

(ii) Claimant's addressees must share a more or less equivalent knowledge of $O_a$ (if a Claimant succeeds in convincing someone that a pink dollar bill bearing the portrait of Gorbachev is good American currency, this would be not forgery but defrauding the mentally incapable).

These requirements being met, there is Downright Forgery when the Claimant claims, in good or bad faith, that $O_b$ is identical with $O_a$, which is known to exist and to be highly valued.

### 4.1.1. *Deliberate false identification*

The Claimant knows that $O_b$ is only a reproduction of $O_a$. Nevertheless, s/he claims, with intent to deceive, that $O_b$ is identical with $O_a$. This is forgery in the narrower sense — offering a copy of the *Mona Lisa* as the original, or putting forged bank notes into circulation.[3]

### 4.1.2. *Naive false identification*

The Claimant is not aware that the two objects are not identical. Thus s/he, in good faith, takes $O_b$ to be the genuine original. This is the case with those tourists who in Florence fetishistically admire outside Palazzo Vecchio the copy of Michelangelo's *David* (without knowing that the original is preserved elsewhere).

### 4.1.3. *Authorial copies*

After completing the object $O_a$, the same author produces in the same manner a perfect double $O_b$, which cannot outwardly be distinguished from $O_b$. Ontologically speaking the two objects are physically and historically distinct, but the author — more or less honestly — believes that from the aesthetic point of view they both have equal value. One may think here of the polemics about the 'forged' pictures by De Chirico, which in the opinion of many critics were painted by De Chirico himself. Such cases provoke a critical questioning of the fetishistic veneration of the artistic original.

### 4.1.4. *Alteration of the original*

A variant of the previous cases occurs when *B* alters $O_a$ to $O_b$. Original manuscripts have been altered, old and rare books have been modified by changing indications of origin and possession, by adding false colophons, by mounting pages from a later edition in order to make up an incomplete copy of a first edition. Paintings and statues are restored is such a way as to alter the work; parts of the body which offend against censorship are covered up or eliminated; parts of the work are removed or a polyptic is separated into its component parts.[4]

Such alterations may be made both in good and in bad faith, depending on whether one believes or does not believe that $O_b$ is still identical with $O_a$, that is, that the object was altered in accordance with the *intentio auctoris*. In actual fact we see as original and authentic ancient works of art which have been substantially altered by the course of time and by human intervention: we have to allow for the loss of limbs, for restoration, and for the fading of colors. In this category belongs the neoclassical dream of a 'white' Greek art, where in fact the statues and temples were originally brightly coloured.

In a certain sense all works of art which have survived from antiquity should be considered as forgeries. But following this line of thought, since any material is subject to physical and chemical alteration, from the very moment of its production, every object should be seen as an instant forgery of itself. To avoid such a paranoiac attitude, our culture has elaborated flexible criteria for deciding about the physical integrity of an object. A book in a bookstore continues to be a brand new exemplar even though opened by many customers, until the moment in which — according to the average taste — it is blatantly worn, dusty or crumpled. In the same vein, there are criteria for deciding when a fresco needs to be restored — even though the contemporary debate on the legitimacy of the restoration of the Sistine Chapel shows us how controversial such criteria are.

The weakness of these criteria provoke, in many case, very paradoxical situations. For instance, from an aesthetic point of view, one usually asserts that a work of art can be recognized as authentically such provided it maintains a basic integrity and that if it is deprived of one of its parts it looses its organic perfection. But from an archaeological and historical point of view one thinks that — even though the same work of art has lost some of its formal features — it is still authentically original provided that its material support — or at least part of it — had remained indiscernibly the same through the years. Thus 'aesthetic authenticity' depends on criteria that are different from those used in order to assert 'archaeological genuineness'. Nevertheless these two notions of authenticity and genuineness interfere in various ways, frequently in an inextricable way. The Parthenon of Athens has lost its colors, a great deal of its original architectural features, and part of its stones: but the remaining ones are — allegedly — the same that the original builders set up. The Parthenon of Nashville, Tennessee, was built according to its Greek model such as it looked at the time of its splendor, it is formally complete, probably colored as the original was intended to be. From the point of view of a purely formal and aesthetic criterion the Greek Parthenon should be considered an alteration or a forgery of the Nashville one. Nevertheless the half temple standing on the Acropolis is considered original while the American copy is just a copy, because temporal priority is considered more important than the presence of given architectural feature. Being original, the Greek Parthenon is also considered more 'beautiful' than the American one.[5]

### 4.2. *Moderate Forgeries*

As for Downright Forgery, we assume that $O_a$ exists or existed in the past, and that the Claimant knows something about it. The addressees know that $O_a$ exists or existed but do not necessarily have clear ideas about it. The Claimant knows that $O_a$ and $O_b$ are different, but decides that in particular circumstances and for particular purposes they are of equal value. S/he does not claim that they are identical but claims that they are interchangeable, since for both the Claimant and the addressees the borderlines between identity and interchangeability are very flexible.

#### 4.2.1. *Confusional enthusiasm*

The Claimant knows that $O_a$ is not identical with $O_b$, the latter having been produced later as a copy, but is not sensitive to questions of authenticity. S/he thinks that the two objects are interchangeable as regards their value and their function and uses or enjoys $O_b$ as if it were $O_a$, thus implicitly advocating their identity.

Roman patricians were aesthetically satisfied with a copy of a Greek statue and asked for a forged signature of the original author. Some tourists in Florence admire the copy of Michelangelo's *David* without being bothered by the fact that it is not the original. At the Getty Museum in Malibu, California, original statues and paintings are inserted in very well reproduced 'original' environments, and many visitors are uninterested in knowing which are the originals and which the copies (see Eco, 1986 b).

4.2.2. *Blatant claim of interchangeability*

This is generally the case with translations, at least from the point of view of the common reader. It was also the case with medieval copies from manuscript to manuscript, where the copyist frequently made deliberate alterations by abbreviating or censoring the original text (still in the belief to be transmitting the 'true' message). In the bookstore of the Museum of the City of New York a facsimile of the bill of sale of Manhattan is sold. In order to look really old, it is scented with old spice. But this Manhattan purchase contract, penned in pseudo-antique characters, is in English, whereas the original was in Dutch.

### 4.3. *Forgery Ex-Nihilo*

Let us rank under this heading (i) works made *à la manière de . . .* , (ii) Apocrypha and Pseudo-epigrapha, (iii) creative forgery.[6]

We must assume (by temporarily suspending any ontological commitment, see 4.1) that $O_a$ does not exist — or, if according to uncertain report it existed in the past, it is by now irremediably lost. The Claimant claims — in good or bad faith — that $O_b$ is identical with $O_a$. In other words, the Claimant falsely attributes $O_b$ to a given author. In order to make this false attribution credible, one must know of a set *a* of different objects ($O_{a1}$, $O_{a2}$, $O_{a3}$ . . .) all produced by an author *A* who is famous and well regarded. From the whole set *a*, an abstract type can be derived, which does not take into account all the features of the individual members of *a* but rather displays a sort of generative rule and is assumed to be the description of the way in which *A* produced every member of *a* (style, type of material used, etc.). Since $O_b$ looks as if it has been produced according to this type, it is then claimed that $O_b$ is a previously unknown product of *A*. When such an imitation *ex-nihilo* is openly admitted to be so — frequently as homage or parody — one speaks of a work made *à la manière de . . ..*

4.3.1. *Diplomatic forgery*

In this case the Claimant coincides with the Author *B* and there are two possibilities: (i) the Claimant knows that $O_a$ never existed, (ii) the Claimant believes in good faith that $O_a$ existed but knows that it is irremediably lost. In both cases, since the Claimant knows that $O_b$ is a brand new production, s/he also knows that $O_b$ is not identical with $O_a$. However the Claimant assumes that $O_b$ can fulfill all the functions performed by $O_a$, and consequently presents $O_b$ as if it were the authentic $O_a$.

Whereas a historical forgery refers to a formally authentic charter, which contains false or invented information (as with an authentic confirmation of a false privilege), the diplomatic forgery offers a false confirmation of supposedly authentic privileges. Examples of this are the forged charters produced by medieval monks who wished to antedate the property claims of their monastery. We can assume that they did so because they strongly believed that their monastery had once genuinely received such confirmations. Medieval authors privileged tradition over documents and had a different notion of authenticity. The only form of credible document they possessed was the traditional notice itself. They could only rely on the testimony of the past, and this past had only vague chronological coordinates. Le Goff (1964, pp. 397—402) has observed that the form taken by medieval knowledge is that of folklore: "La preuve de vérité, à l'époque féodale, c'est l'existence 'de toute éternité'". Le Goff adduces a legal dispute of 1252 between the serfs of the chapter of Notre Dame de Paris in Orly and the canons. The canons based their claim to the payment of tithes on the fact that the fama proved it; the oldest inhabitant of the region was questioned on the subject and he replied that it had been so "a tempore a quo non extat memoria". Another witness, the archdeacon John, said that he had seen old charters in the chapter-house which confirmed the custom, and that the canons regarded these charters as authentic because of their script. No one thought it necessary to prove the existence of these charters, let alone investigate their contents; the report that they had existed for centuries was sufficient. In such a culture it was considered perfectly fair to provide a fake document in order to testify a 'true' tradition.

#### 4.3.2. *Deliberate ex-nihilo forgery*

The Claimant knows that $O_a$ does not exist. If the Claimant coincides with the Author *B*, than the Claimant knows that $O_b$ is of recent manufacture. In any case s/he cannot believe that $O_a$ and $O_b$ are the same. Nevertheless s/he claims, fully aware that s/he is not entitled to do so, that the two objects — one real and one imaginary — are identical or that $O_b$ is genuine, and does so with the intention to deceive. This is the case with modern charter forgeries, with many fake paintings (see the fake Vermeer painted in this century by van Meegeren), with forged family trees intended to demonstrate an otherwise unprovable genealogy and with deliberately produced apocryphal writings (like Hitler's diaries).[7]

It is also the case of the thirteenth century poem *De vetula*, which was immediately ascribed to Ovid. One may suppose that the persons who brought the *Corpus Dionysianum* into circulation in the ninth century and ascribed it to a pupil of St. Paul were well aware that the work was composed much later; nevertheless they decided to credit it to an unquestionable authority. Slightly similar to the case listed in 4.1.3. is the phenomenon of authorial stylistic forgeries, as when a painter, famous for his works of the twenties, paints in the fifties a work which looks like an unheard-of masterpiece of the early period.

#### 4.3.3. *False ascription in error*

The Claimant does not coincide with *B* and does not know that $O_a$ does not exist. S/he claims in good faith that $O_b$ is identical with $O_a$ (of which she has heard by uncertain report). This is what happened with those who received and took the *Corpus Dionysianum* for a work by a pupil of St. Paul, for those who believed and still believe in the authenticity of the Book of Enoch, and for the Renaissance Neoplatonist who ascribed the *Corpus Hermeticum* not to Hellenistic authors but to a mythical Hermes Trismegistos, who was supposed to have lived before Plato in the time of the Egyptians and presumably to be identified with Moses. In this century, Heidegger wrote a commentary on a speculative grammar which he ascribed to Duns Scotus, though it was shown shortly afterwards that the work was composed by Thomas of Erfurt. This seems also to be the case with the ascription of *On the Sublime* to Longinus.[8]

## 5. THE FAKE AS A FAKE SIGN

The above typology suggests some interesting semiotic problems.

First of all, is a fake a sign? Let us first consider the cases of Downright Forgeries (where $O_a$ exists somewhere).

If a sign is — according to Peirce (1934, 2.228) — "something which stands to somebody for something in some respects or capacity", then

$O_b$ stands to the Claimant for $O_a$. And if an icon — still according to Peirce (1934, 2.276) — "may represent its object mainly by its similarity", then $O_b$ is an icon of $O_a$.

$O_b$ succeeds in being mistaken for $O_a$ in so far as it reproduces the whole of $O_a$'s properties. Morris (1946, 1.7) suggests that a "completely iconic sign" is no longer a sign because "it would be itself a denotatum". This means that a possibly completely iconic sign of myself would be the same as myself. In other words complete iconism coincides with indiscernibility or identity, and a possible definition of identity is 'complete iconism'.

But in forgery there is only an alleged identity: $O_b$ can have all the properties of $O_a$ except that of being $O_a$ itself and of standing at the same moment in the same place as $O_a$. Being incompletely iconic, can $O_b$ be taken as a sign of $O_a$?

If so, it would be a rather curious kind of sign: it would succeed in being a sign insofar as nobody takes it as a sign and everybody mistakes it with for its potential denotatum. As soon as one recognizes it as a sign, $O_b$ becomes something similar to $O_a$ — but cannot be any longer confused with $O_a$. In fact, facsimiles are iconic signs but are not fakes.

How are we to define a sign that works as such only if and when it is mistaken for its own denotatum? The only way to define it, is to call it a fake. A peculiar situation, indeed. What kind of semiotic object is a fake?

The question that the Claimant asks when facing $O_b$ is not "what does it mean?" but rather "what is it?" (and the answer which produces a false identification is: "It is $O_a$"). $O_b$ is taken as the same as $O_a$ because it is, or looks like, an icon of $O_a$.

In Peircean terms, an icon is not yet a sign. As a mere image it is a Firstness. Only iconic representamens or hypoicons are signs, that is, instances of Thirdness. Although this point is in Peirce rather controversial, we can understand the above difference in the sense that a mere icon is not interpretable as a sign. Obviously $O_b$, in order to be recognized as similar to $O_a$, must be perceptually interpreted, but as soon as the Claimant perceives it, s/he identifies it as $O_a$. This is a case of *perceptual misunderstanding*.

There is a semiotic process which leads to the perceptual recognition of a given uttered sound as a certain word. If someone utters 'fip' and the addressee understands 'fi:p' certainly the addressee mistakes 'fip' with a token of the lexical type 'fi:p'. But we can hardly say that the uttered 'fip' was a sign for the intended 'fi:p'. The whole story concerns a phonetic muddle or, in so far as both utterances are words, an expression-substance to expression-substance mistake. In the same sense when $O_b$ is mistaken, for reasons of similarity, for a token $O_a$ (and in case of downright forgeries $O_a$ is a token which is the type of itself) we are facing a phenomenon of expression-to-expression misunderstanding.

There are cases in semiosis in which one is more interested in the physical features of a token expression than with its content. For instance when one hears a sentence and is more interested in ascertaining if it was uttered by a certain person than in interpreting its meaning; or when, in order to identify the social status of the speakers, the hearers are more interested in their accent than in the propositional content of the sentence they are uttering.

Likewise, in false identification one is mainly concerned with expressions. Expressions can be forged. Signs (as functions correlating an expression to a content) can at most be misinterpreted.

Let us recall the distinction made by Goodman (1968, pp. 99 ff) between 'autographic' and 'allographic' arts, Peirce's distinction between legisign, sinsign and qualisign (1934, 2.243 ff) and our own previous treatment of replicas (Eco 1976, pp. 178 ff). There are (i) signs whose tokens can be indefinitely produced according to their type (books or musical scores), (ii) signs whose tokens, even though produced according to a type, possess a certain quality of material uniqueness (two flags of the same nation can be distinguished on the grounds of their glorious age), and (iii) signs whose token is their type (like autographic works of visual arts).

From this point of view we are obliged to draw a straightforward distinction between different types of forgery. Let us mainly consider Downright Forgeries and Forgeries Ex-Nihilo (it will be evident in which sense Moderate Forgeries stand in between).

Downright Forgeries only affect signs (ii) and (iii). It is impossible to produce a fake *Hamlet* unless by making a different tragedy or by editing a detectable, censored version of it. It is possible to produce a forgery of its first folio edition because in this case what is forged is not the work of Shakespeare but that of the original printer.

Downright Forgeries are not signs: they are only expressions which look like other expressions — and they can become signs only if we take them as facsimiles.

On the contrary, it seems that phenomena of Forgery Ex-Nihilo are more semioesically complicated. It is certainly possible to claim that a statue $O_b$ is indiscernibly the same as the legendary statue $O_a$ by a great Greek artist (same stone, same shape, same original connection with the hands of its author); but it is also possible to attribute a written document $O_b$ to an author $A$ without paying attention to its expression substance. Before Aquinas, a Latin text, known to be translated from an Arab version, *De Causis*, was attributed to Aristotle. Nobody falsely identified either a given parchment nor a given specimen of handwriting (because it was known that the alleged original object was in Greek). It was *the content* that was (erroneously) thought to be Aristotelian.

In such cases, $O_b$ was first seen as a sign of something in order to recognize this something as absolutely interchangeable with $O_a$ (in the sense examined in 4.2.2).

In Downright Forgeries (and in the case of autographic arts) the Claimant makes a claim about the authenticity, or genuineness or originality of the expression. In Forgeries Ex-Nihilo (which concern both autographic and allographic arts) the Claimant's claim can concern either the expression or the content.

In Downright Forgeries, the Claimant — by virtue of a perceptual misunderstanding concerning two expression substances — believes that $O_b$ is the same as the allegedly authentic $O_a$. In the second case, the Claimant — in order to identify $O_b$ is the expression of a given content which in itself is the same as the genuine and authentic content of the legendary allographic expression $O_a$.[9]

In both cases, however, one feels something uncanny. A naive approach to fakes and forgeries makes one believe that the problem with fakes is to take for granted or to challenge the fact that something is the same as an allegedly authentic object. However after a more accurate inspection it seems that the real problem is to decide what does one mean by 'authentic object'. Ironically, the problem with fakes is not whether $O_b$ is or is not a fake, but rather whether $O_a$ is authentic or not, and on which ground such a decision can be made.

It seems that the crucial problem of a semiotics of fakes is not the one of a typology of the mistakes of the Claimant, but rather of a list of the criteria by which the Judge decides whether the Claimant is right or not.

## 6. CRITERIA FOR ACKNOWLEDGING AUTHENTICITY

The task of the Judge (if any) is to verify or falsify the claim of identity made by the Claimant. The Judge can basically face two alternatives:

(i) *Downright Forgeries* — $O_a$ is largely known to exist, and the Judge has only to prove that $O_b$ is not identical with it. In order to do so the Judge has two further alternatives: either s/he succeeds in putting $O_b$ in place of $O_a$, thus showing that are not indiscernibly identical, or s/he compares the features of $O_b$ with the celebrated and well known features of $O_a$ in order to show that the former cannot be mistaken with the latter.

(ii) *Ex-Nihilo Forgery* — The existence of $O_a$ is a mere matter of tradition and nobody has ever seen it. When there are no reasonable proofs of the existence of something one can assume that it probably does not exist or has disappeared. But the newly-found $O_b$ is usually presented by the Claimant as the expected proof of the existence of the $O_a$. In this case the Judge should prove or disprove that the $O_b$ is authentic. If it is authentic then it is identical with the allegedly lost original $O_a$. However the authenticity of something allegedly similar to a lost original can be demonstrated only by proving that $O_b$ is the original.

This case seems more complicated that the first one. In case (i) it seems that — in order to demonstrate the authenticity of $O_b$ — it was enough to show $O_b$ was identical with the original $O_a$ — and the original $O_a$ represented a sort of unchallengeable parameter. In case (ii) there is no parameter. However let us consider case (i) more closely.

A Judge can know beyond doubt that $O_a$ and $O_b$ are not identical only if someone shows a perfect copy — let us say — of the *Mona Lisa* while standing in front of the original in the Louvre and claims that the two objects are indiscernibly identical. But even in this implausible case there would be a shadow of doubt remaining: perhaps $O_b$ is the genuine original, and $O_a$ is a forgery.

We are thus here facing a curious situation. Forgeries are cases of false identification. If the Judge proves that the objects are two and challenges the false claim of identification, the Judge has certainly proved that there was a case of forgery. But s/he has not yet proved which one of the two objects is the original one. It is not sufficient to prove that the identification is impossible. The Judge must provide a *proof of authentication* for the supposed original.

At first glance case (ii) looked more difficult because, in the absence of the presumed original, one should demonstrate the suspected fake is the original. In fact case (i) is far more difficult: when the original is present one should still demonstrate that the original is the original.

It is not sufficient to say that the $O_b$ is a fake because it does not possess all the features of the $O_a$. The method by which the Judge identifies the features of any $O_b$ is the same as that with which the Judge makes a decision about the authenticity of the $O_a$. In other words, in order to say that a reproduction is not the genuine *Mona Lisa*, one must have examined the genuine *Mona Lisa* and confirmed its authenticity with the same techniques as one uses to say that the reproduction differs from the original. Modern philology is not content with the testimony that, let us say, the *Mona Lisa* was hung in the Louvre by Leonardo as soon as he had completed it. This claim would have to be proved by documents, and this in turn would raise the question of the documents' authenticity.

In order to prove that an $O_b$ is a fake, a Judge must prove that the corresponding $O_a$ is authentic. Thus the Judge must examine the presumably genuine painting *as if it were a document*, in order to decide whether its material and formal features allow the assumption that it was authentically painted by Leonardo.

Modern scholarship proceeds therefore from the following assumptions:

(i) A document confirms a traditional belief and not the other way round.

(ii) Documents can be: (a) objects produced with an explicit intention of communication (manuscripts, books, gravestones, inscriptions, etc.), where one can recognize an expression and a content (or an intentional meaning); (b) objects which were not primarily intended to communicate (such as prehistoric finds, objects of everyday used in archaic and primitive cultures) and which are interpreted as signs, symptoms, traces of past events; (c) objects produced with an explicit intention of communicating $x$, but taken as non-intentional symptoms of $y$ — $y$ being the result of an inference about their origin and their authenticity.

(iii) Authentic means historically original. To prove that an object is original means considering it *as a sign of its own origins.*

Thus if a fake is not a sign, for modern philology the original, in order to be compared with its fake copy, must be approached as a sign. False identification is a semiotic web of misunderstandings and deliberate lies, while any effort to make a 'correct' authentication is clear case of semiotic interpretation or of *abduction.*

### *6.1. Proofs Through Material Support*

A document is a fake if its material support does not date back to the time of its alleged origin. This kind of proof is a rather recent one.

Greek philosophers looking for the sources of an older of Oriental wisdom rarely had any chance of dealing with original texts in their original language. The medieval translators generally worked with manuscripts which stood at a considerable distance from the archetype. As for the artistic marvels of antiquity, people in the Medieval age only knew either crumbling ruins or vague rumors about unknown places. The judgements passed in the early Middle Ages on whether a document produced in evidence in a lawsuit was genuine or not were at best restricted to investigating the authenticity of the seal.

Even during the Renaissance, the same scholars who started studying Greek and Hebrew, when the first manuscript of the *Corpus Hermeticum* was brought to Florence and was attributed to a very remote author, did not wonder at the fact that the sole physical evidence they had — the manuscript — dated to the fourteenth century.

Nowadays there are recognized physical or chemical techniques for determining the age and the nature of a medium (parchment, paper, linen, wood, etc.) and these means are considered fairly 'objective'. In these cases, the material support — which is an instance of the substance of the expression — must be examined in its physical structure, that is, as a form (see Eco 1976, 3.7.4 on the "overcoding of the expression"). In fact the generic notion of material support must be further analyzed into subsystems and subsystems of subsystems. For instance, in a manuscript, writing is the substitute for the linguistic substance, inking is the support of the graphematic manifestation (to be seen as a form), the physico-chemical features of the parchment are the support of its formal qualities, and so on and so forth. In a painting, brush strokes are the support of the iconic manifestation, but they become in their turn the formal manifestation of a pigmentary support, etc.

### 6.2. *Proofs Through Text Linear Manifestation*

The text linear manifestation of a document must conform to the normative rules of writing, painting, sculpting and so on holding at the moment of its alleged production. The text linear manifestation of a given document must thus be compared with everything known about the system of the form of the expression in a given period — as well as with what is known of the personal style of the alleged author.

Augustine, Abelard and Aquinas were confronted with the problem of determining the credibility of a text from its linguistic characteristics. However, Augustine, whose knowledge of Greek was minimal and who knew no Hebrew, advises in a passage on *emendatio* that when dealing with Biblical texts one should compare a number of different Latin translations, in order to be able to conjecture the 'correct' translation of a text. He sought to establish a 'good' text, not an 'original' text, and he rejected the idea of using the Hebrew text because he regarded this as having been falsified by the Jews. As Marrou (1958, pp. 432—434) remarks, "ici réapparait le grammaticus antique . . . Aucun de ses commentaires ne suppose un effort préliminaire pour établir critiquement le texte . . . Aucun travail préparatoire, nulle analyse de la tradition manuscrite, de la valeur précise des différents témoins, de leur rapports, de leur filiation: saint Augustin se contente de juxtaposer sur la table le plus grand nombre de manuscrits, de prendre en considération dans son commentaire le plus grand nombre de variantes". The last word lay not with philology but with the honest desire to interpret and the belief in the validity of the knowledge so transmitted. Only in the course of the thirteenth century did scholars begin to ask converted Jews in order to obtain information on the Hebrew original (Chenu 1950, pp. 117—125 and 206).

St. Thomas paid attention to the *usus* (by which he understood the lexical usage of the period to which a given text refers, cf. *Summa Theol.* I, 29, 2 ad 1). By considering the *modus loquendi* he argued that in particular passage Dionysius or Augustine used particular words because they were following the practice of the Platonist. Abelard in *Sic et Non* argued that one should mistrust an allegedly authentic text where words are used with unusual meanings, and that textual corruption can be a sign of forgery. But practice fluctuated, at least until Petrarch and the proto-humanists.

The first example of philological analysis of the form of expression is provided in the fifteenth century by Lorenzo Valla (*De falso credita et ementita Constantini donatione declamatio*, xiii) when he shows that the use of certain linguistic expressions was absolutely implausible at the beginning of the fourth century A.D. Likewise Isaac Casaubon at the beginning of the the sixteenth century (*De rebus sacris et ecclesiasticis exercitationes* XIV) proved that the *Corpus Hermeticum* was not a Greek translation of an ancient Egyptian text because it does not bear any trace of Egyptian idioms. Modern philologists demonstrate that the Hermetic *Asclepius* was not translated, as assumed before, by Marius Victorinus because Victorinus in all his texts consistently put *etenim* at the beginning of the sentence, while in the *Asclepius* this word appears in the second position in twenty one cases out of twenty five (Nock, 1945).

Today we resort to many paleographic, grammatical, iconographic and stylistic criteria based upon a vast knowledge of our cultural heritage. A typical example of modern technique for attributing paintings was that of Morelli (see Ginzburg, 1979) based upon the most marginal features, such as the way of representing fingernails or the ear lobe. These criteria are not irrefutable but represent a satisfactory basis for philological inferences.

### *6.3. Proofs through Content*

For such proofs it is necessary to determine whether the conceptual categories, taxonomies, modes of argumentation, iconological schemes and so on are coherent with the semantic structure (the form of the content) of the cultural milieu of the alleged authors — as well as with the personal conceptual style of these authors (extrapolated from their other works).

Abelard tried to establish when the meaning of words varies with particular authors and recommended — as had Augustine in *De Doctrina Christiana* — the use of contextual analysis. But this principle is restricted by the parallel recommendation to give preference to the more important authority in case of doubt.

When Aquinas questioned the false ascription of *De Elementatio theologica* by Proclus (which had just been translated by William of Moerbecke) and by showing that the former had the same Neoplatonic content as the latter. This philosophical attitude was undoubtedly very mature, but Aquinas usually did not ask whether people thought and wrote according to the world-view of their times, but rather whether it was 'correct' to think and write in such a way and therefore whether the text could be ascribed to doctrinal authorities who were never wrong.

Aquinas repeatedly used the term *authenticus*, but for him (as for the Middle Ages in general) the word did not mean 'original' but 'true'.

*Authenticus* denotes the value, the authority, the credibility of a text, not its origin: of a passage in *De causis* it is said "ideo in hac materia non est authenticus" (*II Sent.* 18, 2,2 ad 2), but the reason is that here the text cannot be reconciled with Aristotle.

As Thurot (1869, pp. 103—104) says, "en expliquant leur texte les glossateurs ne cherchent pas à entendre la pensée de leur auteur, mais à enseigner la science elle-même que l'on supposait y être contenue. Un auteur *authentique*, comme on disait alors, ne peut ni se tromper, ni se contredire, ni suivre un plan défectueux, ni être en désaccord avec un autre auteur authentique."

By contrast, one can find a modern approach to the content form in Lorenzo Valla, when he shows that a Roman emperor like Constantine could not have thought what the *Constitutum* (falsely attributed to him) said. Likewise, Isaac Casaubon's argument against the antiquity of the *Corpus Hermeticum* is that if in these texts were to be found echoes of Christian ideas then they had been written in the first centuries of our era.

However, even today such criteria (though based upon an adequate knowledge of the world-views prevailing in different historical periods) are naturally dependent to a large extent on suppositions and abductions which are open to challenge.

### *6.4. Proofs through External Evidences (Referent)*

According to this criterion, a document is a fake if the external facts reported by it could not have been known at the time of its production. In order to apply this criterion one must display adequate historical knowledge but must also hold that it is implausible that the alleged ancient author had the gift of prophecy. Before Casaubon, Ficino and Pico della Mirandola had read the *Corpus Hermeticum* by breaching this principle: they considered the Hermetic writings as divinely inspired simply because they 'anticipated' Christian conceptions.

In the Middle Ages, some opponents of the Donation of Constantine tried to reconstruct the facts and reject the test as apocryphal because it contradicted what they knew about the past. In a letter to Frederick Barbarossa in 1152, Wezel, a follower of Arnold of Brescia, argued that the Donation was a *mendacium* because it contradicted other witnesses of the period, which showed that Constantine had been baptized under other circumstances and at a different time. The criticism became more rigorous in the early humanistic era: for example, in the *Liber dialogorum hierarchie subcelestis* of 1388 and in the *De concordantia Catholica* by Nicholas of Cusa, the author tries to establish historical truth by careful evaluation of all the sources.

Lorenzo Valla displayed more indisputable historical proofs: for instance, he proved that the Donation speaks of Constantinople as a patriarchate when at the supposed time of composition Constantinople did not exist under that name and was not yet a patriarchate.

Recent study of an alleged exchange of correspondence between Churchill and Mussolini has shown that, despite the genuineness of the paper used, the correspondence must be rejected and considered as a forgery because it contains evident factual contradictions. One letter is dated from a house in which Churchill had not at that time lived in for years, another deals with events which occurred after the date of the letter.

### 6.5. *The Intention of the Author*

According to contemporary text semiotics one should consider apart all the inferences about the *intentio auctoris*. The intention of the author, as the Sender of the examined message, is different from the content. It can be manifested through different strategies at many levels of the examined message.

The interpretation of a text aims at outlining an ideal image of a Model Author. By knowing the operative strategies of the alleged author, philologists can make abductions about the correspondence between the authorial intention displayed by $O_b$ and the intentions attributed to the alleged $A$. This way of reading is highly conjectural but can be used for purposes of authentication (see Eco, 1979 and 1986a).

## 7. CONCLUSIONS

It thus seems that our modern culture has outlined 'satisfactory' criteria for proving authenticity and for falsifying false identifications. However all the above criteria seem very useful only when a Judge is faced with 'imperfect' forgeries. Is there a 'perfect forgery' which defies any philological criterion? Or are there cases in which no external proofs are available while the internal ones are highly arguable?

Let us imagine the following:

In 1921 Picasso asserts that he has painted a portrait of Honorio Bustos Domeq. Fernando Pessoa writes that he has seen the portrait and praises it as the greatest masterpiece ever produced by Picasso. Many critics look for the portrait but Picasso says that it has been stolen.

In 1945, Salvador Dali announces that he has rediscovered this portrait in Perpignan. Picasso formally recognizes the portrait as his original work. The painting is sold to the Museum of Modern Art, under the title 'Pablo Picasso: Portrait of Bustos Domeq, 1921'.

In 1950, Jorge Luis Borges writes an essay ('El Omega de Pablo') in which he maintains that:

1. Picasso and Pessoa lied because nobody in 1921 painted any portrait of Domeq.
2. In any case, no Domeq could have been portrayed in 1921 because such a character was invented by Borges and Bioy Casares during the forties.
3. Picasso actually painted the portrait in 1945 and falsely dated it 1921.
4. Dali stole the portrait and copied it (masterfully). Immediately afterwards he destroyed the original.
5. Obviously the 1945 Picasso was perfectly imitating the style of the early Picasso and Dali's copy was indistinguishable from the original. Both Picasso and Dali used canvas and colors produced in 1921.
6. Therefore the work exposed in New York is the deliberate authorial forgery of a deliberate forgery of a historical forgery (which mendaciously portrayed a non-existent person).

In 1986 there is found an unpublished text of Raymond Queneau asserting that:

1. Bustos Domeq really existed, except that his real name was Schmidt. Alice Toklas in 1921 maliciously introduced him to Braque as Domeq and Braque portrayed him under this name (in good faith) imitating the style of Picasso (in bad faith).
2. Domeq-Schmidt died during the blanket bombing of Dresden and all his identity papers were destroyed in those circumstances.

3. Dali really rediscovered the portrait in 1945 and copied it. Later, he destroyed the original. A week later Picasso made a copy of Dali's copy, then the copy by Dali was destroyed. The portrait sold to the MOMA is a fake painted by Picasso imitating a fake painted by Dali imitating a fake painted by Braque.
4. He (Queneau) has learnt all this for sure from the discoverer of Hitler's diaries.

All the individuals involved in this story are by now dead. The only object we have at our disposal is that hanging in the MOMA.

It is evident that none of the philological criteria listed in Section 6 can help us in ascertaining the truth. Even though it is possible that a perfect connoisseur can distinguish some imponderable differences between the hand of Dali and the hand of Picasso, or between the two hands of Picasso in different historical periods, any assertion of this kind could be challenged by other experts.

Such a story is not as paradoxical as it might seem. We are still wondering if the author of the Iliad was the same as the author of the Odyssey, if one of them (at least) was Homer and if Homer was a single person.

The current notion of fake presupposes a 'true' original with which the fake should be compared. But we have seen that every criterion for ascertaining whether something is the fake of an original coincides with the criteria for ascertaining whether the original is authentic. Thus the original cannot be used as a parameter for unmasking its forgeries unless we blindly take for granted that what is presented to us as the original is unchallengeable so (but this would contrast with any philological criterion).

Proofs through material support tell us that a document is a fake if its material support does not date back to the time of its alleged origin. Such a test can clearly prove that a canvas produced with a mechanical loom cannot have been painted during the sixteenth century, but cannot prove that a canvas produced in the sixteenth century and covered with colors chemically similar to those produced at that time, was really painted during the sixteenth century.

Proofs through Text Linear Manifestation tell us that a text is fake if its text linear manifestation does not conform to the normative rules of writing, painting, sculpting and so on holding at the moment of its alleged production. But the fact that a text meets all these requirements does not prove that its text is original (this proves at most that the forgerer was very skilled).

Proofs through content tell us that a text is a fake if its conceptual categories, taxonomies, modes of argumentation, iconological schemes and so on, are not coherent with the semantic structure (the form of the content) of the cultural milieu of the alleged authors. But there is no way to demonstrate that a text was originally written B. C. only because it does not contain Christian ideas.

Proofs through external evidences tell us that a document is a fake if the external facts reported by it could not have been known at the time of its production. But there is no way to demonstrate that a text which reports events that happened at the time of its alleged production is — for this sole reason — original.

Thus a semiotic approach to fakes shows how theoretically weak are our criteria for deciding about authenticity.

Let us however suppose that there are satisfactory authenticity-tests. If every identification of an original as such presupposes a careful scrutiny of its authenticity, one should test the *Mona Lisa* every time one goes to the Louvre, since without the authenticity test there will be no proof that the *Mona Lisa* seen today is indiscernibly identical with the one seen last week.

Such a test would be necessary for every judgment of identity. As a matter of fact, there is no ontological guarantee that the John I meet today is the same as the John I met yesterday. John undergoes physical (biological) changes much more than a painting or a statue. John can intentionally disguise himself in order to look like Tom.

However in order to recognize John, our parents, husbands, wives and sons every day (as well as in order to decide that the Tour Eiffel I see today is the same as the one I saw last year) we rely on certain instinctive procedures. They prove to be reliable because by using them our species has succeeded in surviving for millions of years and we are world-adapted beings. We never cast these procedures in doubt because it is very rare for a human being or a building to be forged (the rare exceptions to this rule are only interesting subject matter for detective stories or science fiction). But, in principle, John is no more difficult to forge than the *Mona Lisa*; on the contrary it is easier to disguise a person successfully than successfully to copy the *Mona Lisa*.

Objects, documents, bank notes and works of art are frequently forged not because they are particularly easy-to-forge but for mere economic reasons. However, the fact that they *are* so frequently forged obliges us to ask so much about the requirements an original should meet in order to be defined as such — while we do not usually reflect on all other cases of identification.

The reflection on these most commonly forged objects should however tell us how hazardous are our general criteria for identity and how much such concepts as Truth and Falsity, Authentic and Fake, Identity and Difference circularly define each other.

*University of Bologna*

NOTES

* A first version of this paper was presented in September 1986 as the Opening Speech at the Congress on *Falschung im Mittelalters* organized in Münich by *Monumenta Germaniae Historica.* A shorter version was presented in July 1986 at the symposium on *L'à peu près* in Urbino (Centro Internazionale di Linguistica e Semiotica). The present version takes into account the discussion that followed in the course of a seminar on the semiotics of fakes held at the University of Bologna, 1986—87. The present version was ready for printing when I had the chance to see *Faking It. Art and the Politics of Forgery,* by Ian Haywood, New York, Saint Martin Press, 1987. Reference to this book are introduced into the notes.

[1] See also Haywood 1987, pp. 10—18.

[2] Cf. Haywood 1987, Ch. 2 on literary forgeries. In this sense every novel which is presented as the transcription of an original manuscript, a collection of letters, and so on, could be intended as a form of historical forgery. But on this line of thought, every novel, in so far as it is presented as a report about real events, would be a historical forgery. What usually prevents novels from being so, is the whole series of more or less perceptible 'genre signals' that transform any pretended assertion of authenticity into a tongue-in-check statement.

[3] See Haywood 1987, pp. 91 ff, the question of the fake fossilised remains.

[4] See Haywood, p. 42 ff. on editorial interference.

[5] Goodman (1968) says: "A forgery of a work of art is an object falsely purporting to have the history of production for the (or an) original of the work" (p. 122). Thus the Parthenon of Nashville would be a forgery (or at least a mere copy) because it has not the same story than the one of Athen. But this wouldn't be sufficient, since Goodman admits that architecture can be considered as an allographic art. Given a precise plan (type) of the Empire State Building, there would be no difference between a token of that type built in midtown Manhattan and another token built in the Nevada desert. In fact the Greek Parthenon is not only 'beautiful' because of its proportions and of other formal qualities (severely altered in the course of the last two thousand years) but because of its natural and cultural environment, its location on the top of a hill, all the literary and historical connotations it suggests.

[6] See Haywood (Ch. 1) for apocrypha and creative forgeries.

[7] On van Meegeren see Haywood, Ch. 5, Goodman (1968), Barbieri (1987) and the bibliography in Haywood.

[8] See the chapter devoted by Haywood to the Schliemann case as a complex web of different cases of ex-nihilo forgery. "Not only had Schliemann not uncovered Priam's fabled city (but a much earlier one) — but it has recently been revealed that Schliemann's discovery of the fabulous treasure which became world famous was a hoax . . . . Most of the treasure was genuine in the sense of being genuinely old . . . . The treasure was a forgery because its provenance was false. Schliemann even inserted the fictitious tale of discovery into his own diary . . . . The parts were genuine but the whole was fictional. Schliemann forged authentication and invented a context" (pp. 91—92).

[9] If an Author *B* copies a book $O_a$ and says "this is $O_a$, made by Author *A*", then she says something true. If on the contrary the same Author *B* copies a painting or a statue $O_a$ and says "this is $O_a$, made by Author *A*", then s/he certainly says something false (if both say that $O_b$ is their own work, they are guilty of plagiarism). But is it true that an Author *B* who has masterly copied an $O_a$ and presents it as his/her own work is asserting something blatantly false? Autographic works being their own type, to imitate them perfectly provides the imitation with a proper aesthetic quality. The same happens with Ex-Nihilo Forgeries, for instance when an Author *B* produces a painting *à la manière de* . . . . The *Disciples at Emmaus* painted by van Meegeren — and falsely attributed to Vermeer — was undoubtedly a forgery, from the ethical and legal point of view (at least once van Meegeren claimed that it was made by Vermeer). But as a work of art, it was a genuine 'good' painting. If van Meegeren presented it as a homage, it would have been praised as a splendid post-modern endeavor. On such a web of contrasting criteria, see Haywood, Ch. 5 and this quotation from Frank Arnau (*The Thousand Years of Deception in Art and Antiques*, London, Cape, 1961, p. 45): "the boundaries between permissible and impermissible, imitation, stylistic plagiarism, copy, replica and forgery remain nebulous".

## REFERENCES


Barbieri, Daniele: 1987, 'Is reality a fake?', *Versus* **46**.

Chenu, M. D.: 1950, *Introduction à l'étude de Saint Thomas d'Aquin*, Paris: Vrin.

De Leo, Pietro: 1874, *Ricerche sui falsi medievali*, Reggio Calabria: Editori Meridionali Riuniti.

Eco, Umberto: 1986, *Travels in Hyperreality*, New York: Harcourt Brace Jovanovich.

Eco, Umberto: 1979, *The Role of the Reader*, Bloomington: Indiana U.P.

Eco, Umberto: 1984, *Semiotics and the Philosophy of Language*, Bloomington: Indiana U.P.

Ginzburg, Carlo: 1979, 'Morelli, Freud and Sherlock Holmes', in Eco, U. and Sebeok, T. A. (eds.), *The sign of three*, Bloomington: Indiana U.P.

Goodman, Nelson: 1968, *Languages of Art*, New York: Bobbs-Merril.

Hermes, Trismegistus: 1983, *Corpus Hermeticum* (4 vol. Editing by A. D. Nock and translated by A.-J. Festugière), Paris: Société d'édition 'Les Belles Lettres' (6e éd.).

Hintikka, Jaakko: 1969, 'On the logic of perception', in: *Models for Modalities*, Dordrecht: Reidel.

Le Goff, Jacques: 1964, *La civilisation de l'Occident médieval*, Paris: Arthaud.
Marrou, Henri-Irené: 1958, *Saint Augustin et la fin de la culture antique*, Paris: Vrin.
Peirce, Charles Sanders: 1934—48, *Collected Papers* (4 vol.), Cambridge: Harvard U.P.
Prierto, Luis: 1975, *Pertinence et pratique*, Paris: Minuit.
Thurot, C.: 1869, *Extraits de divers manuscrits latins pour servir à l'histoire des doctrines grammaticales du Moyen Age*, Paris.

# APPENDIX: SYMPOSIUM PROGRAM AND LIST OF CONTRIBUTORS

*Understanding Origins:*
*Ideas on the Genesis of Life, Mind and Society*
*An International Symposium held at Stanford University,*
*September 13—16, 1987.*

MONDAY MORNING

*Opening Lecture*

René Girard (Stanford University):
'Origins: A View from the Literature'

*Violence and the Origin of Social Order*

*Presenter*:
Andrew McKenna (Loyola University, Chicago)

*Discussants*:
Paul Dumouchel (University of Waterloo, Ontario)
Paisley Livingston (McGill University)
Renato Rosaldo (Stanford University)

MONDAY AFTERNOON

*Symbols, Texts and the Origin of Money*

*Presenter*:
André Orléan (CREA, Ecole Polytechnique, Paris)

*Discussants*:
Daniel de Coppet (EHSS, Paris)
Jean-Joseph Goux (Brown University)



*Francisco J. Varela and Jean-Pierre Dupuy (eds), Understanding Origins, 305—308.*

TUESDAY MORNING

*Evolution and the Diversity of Life*

*Presenter*:
Stuart Kauffman (Univ. of Pensylvannia, Philadelphia)

*Discussants*:
Daniel Brooks (Univ. of British Columbia)
John Dupré (Stanford University)
Brian Goodwin (Open University, Milton Keynes, UK)
Susan Oyama (CUNY, New York)

TUESDAY AFTERNOON

*Special Lectures:*

Antonio Lazcano (National Polytechnic Institute, Mexico):
'The Origin of Life'
Umberto Eco (Univ. of Bologna):
'The Real and the Copy'

WEDNESDAY MORNING

*Perception and the Origin of Cognition*

*Presenter*:
Francisco Varela (CREA, Ecole Polytechnique, Paris)

*Discussants*:
Jerome Feldman (Univ. of Rochester)
Walter Freeman (Univ. of California at Berkeley)
Christine Skarda (CREA, Ecole Polytechnique, Paris)
Gunther Stent (Univ. of California at Berkeley)

WEDNESDAY AFTERNOON

*Language and the Origin of Meaning*

*Presenter*:
Thomas Bever (Univ. of Rochester)

*Discussants*:
Alvin Liberman (Univ. of Connecticut)
Massimo Piatelli-Palmerini (MIT)
Terry Winograd (Stanford University)

*Invited Discussants*:
Daniel Andler (CREA, Ecole Polytechnique, Paris)
Jean-Marie Apostolides (Stanford University)
Hubert Dreyfus (Univ. of California at Berkeley)
Jean-Pierre Dupuy (CREA, Ecole Polytechnique, Paris)
Marcus Feldman (Stanford University)
William Thompson (Lindisfarne Institute, Zurich)

## LIST OF CONTRIBUTORS

JEAN-PIERRE DUPUY, CREA, Ecole Polytechnique, 1 rue Descartes, 75005 Paris, France

FRANCISCO VARELA, CREA, Ecole Polytechnique, 1 rue Descartes, 75005 Paris, France

RENÉ GIRARD, Dept. of French and Italian, Stanford, Stanford, CA, USA.

ANDREW MCKENNA, Dept. Modern Languages and Literature, Loyola University of Chicago, 6525 North Sheridan Rd., Chicago, IL 60626, USA.

PAUL DUMOUCHEL, Dept. of Philosophy, Université de Québec à Montreal, Sucursal A, CP 8888, Montreal, PQ H3C3P8, Canada.

PAISLEY LIVINGSTON, Dept. of English, McGill University, 853 Sherbrooke Street West, Montrol, PQ H3A 2T6, Canada.

ANDRÉ ORLÉAN, CREA Ecole Polytechnique, 1 rue Descartes, 75005 Paris, France.

STUART KAUFFMAN, Dept. of Physiology and Biophysics, Univ. of Pennsylvania Medical School, Philadelphia, PA, USA.

JOHN DUPRÉ, Dept. of Philosophy, Stanford University, Stanford, CA, USA.

DANIEL BROOKS, Dept. of Zoology, University of Toronto, Toronto, Ontario, M5S 1A1, Canada.

BRIAN GOODWIN, Dept. of Biology, The Open University, Walton Hall, Milton Keynes MK7 6AA, England.

SUSAN OYAMA, 924 West End Av. #44, New York, NY 10025, USA.

CHRISTINE SKARDA, c/o Freeman, Dept. of Physiology-Anatomy, Univ. of California at Berkeley, Berkeley, CA, USA.

UMBERTO ECO, Dept. of Communications, Universitá di Bologna, Bologna, Italy.

# INDEX OF NAMES

## INDEX OF SUBJECTS

# Boston Studies in the Philosophy of Science

*Editor:* Robert S. Cohen, *Boston University*

1. M.W. Wartofsky (ed.): *Proceedings of the Boston Colloquium for the Philosophy of Science, 1961/1962.* [Synthese Library 6] 1963
ISBN 90-277-0021-4
2. R.S. Cohen and M.W. Wartofsky (eds.): *Proceedings of the Boston Colloquium for the Philosophy of Science, 1962/1964.* In Honor of P. Frank. [Synthese Library 10] 1965 ISBN 90-277-9004-0
3. R.S. Cohen and M.W. Wartofsky (eds.): *Proceedings of the Boston Colloquium for the Philosophy of Science, 1964/1966.* In Memory of Norwood Russell Hanson. [Synthese Library 14] 1967 ISBN 90-277-0013-3
4. R.S. Cohen and M.W. Wartofsky (eds.): *Proceedings of the Boston Colloquium for the Philosophy of Science, 1966/1968.* [Synthese Library 18] 1969
ISBN 90-277-0014-1
5. R.S. Cohen and M.W. Wartofsky (eds.): *Proceedings of the Boston Colloquium for the Philosophy of Science, 1966/1968.* [Synthese Library 19] 1969
ISBN 90-277-0015-X
6. R.S. Cohen and R.J. Seeger (eds.): *Ernst Mach, Physicist and Philosopher.* [Synthese Library 27] 1970 ISBN 90-277-0016-8
7. M. Čapek: *Bergson and Modern Physics.* A Reinterpretation and Re-evaluation. [Synthese Library 37] 1971 ISBN 90-277-0186-5
8. R.C. Buck and R.S. Cohen (eds.): *PSA 1970.* Proceedings of the 2nd Biennial Meeting of the Philosophy and Science Association (Boston, Fall 1970). In Memory of Rudolf Carnap. [Synthese Library 39] 1971
ISBN 90-277-0187-3; Pb 90-277-0309-4
9. A.A. Zinov'ev: *Foundations of the Logical Theory of Scientific Knowledge (Complex Logic).* Translated from Russian. Revised and enlarged English Edition, with an Appendix by G.A. Smirnov, E.A. Sidorenko, A.M. Fedina and L.A. Bobrova. [Synthese Library 46] 1973
ISBN 90-277-0193-8; Pb 90-277-0324-8
10. L. Tondl: *Scientific Procedures.* A Contribution Concerning the Methodological Problems of Scientific Concepts and Scientific Explanation.Translated from Czech by D. Short. [Synthese Library 47] 1973
ISBN 90-277-0147-4; Pb 90-277-0323-X
11. R.J. Seeger and R.S. Cohen (eds.): *Philosophical Foundations of Science.* Proceedings of Section L, 1969, American Association for the Advancement of Science. [Synthese Library 58] 1974 ISBN 90-277-0390-6; Pb 90-277-0376-0
12. A. Grünbaum: *Philosophical Problems of Space and Times.* 2nd enlarged ed. [Synthese Library 55] 1973 ISBN 90-277-0357-4; Pb 90-277-0358-2

13. R.S. Cohen and M.W. Wartofsky (eds.): *Logical and Epistemological Studies in Contemporary Physics.* Proceedings of the Boston Colloquium for the Philosophy of Science, 1969/72, Part I. [Synthese Library 59] 1974
ISBN 90-277-0391-4; Pb 90-277-0377-9
14. R.S. Cohen and M.W. Wartofsky (eds.): *Methodological and Historical Essays in the Natural and Social Sciences.* Proceedings of the Boston Colloquium for the Philosophy of Science, 1969/72, Part II. [Synthese Library 60] 1974
ISBN 90-277-0392-2; Pb 90-277-0378-7
15. R.S. Cohen, J.J. Stachel and M.W. Wartofsky (eds.): *For Dirk Struik.* Scientific, Historical and Political Essays in Honor of Dirk J. Struik. [Synthese Library 61] 1974 ISBN 90-277-0393-0; Pb 90-277-0379-5
16. N. Geschwind: *Selected Papers on Language and the Brains.* [Synthese Library 68] 1974 ISBN 90-277-0262-4; Pb 90-277-0263-2
17. B.G. Kuznetsov: *Reason and Being.* Translated from Russian. Edited by C.R. Fawcett and R.S. Cohen. 1987 ISBN 90-277-2181-5
18. P. Mittelstaedt: *Philosophical Problems of Modern Physics.* Translated from the revised 4th German edition by W. Riemer and edited by R.S. Cohen. [Synthese Library 95] 1976 ISBN 90-277-0285-3; Pb 90-277-0506-2
19. H. Mehlberg: *Time, Causality, and the Quantum Theory.* Studies in the Philosophy of Science. Vol. I: *Essay on the Causal Theory of Time.* Vol. II: *Time in a Quantized Universe.* Translated from French. Edited by R.S. Cohen. 1980 Vol. I: ISBN 90-277-0721-9; Pb 90-277-1074-0
Vol. II: ISBN 90-277-1075-9; Pb 90-277-1076-7
20. K.F. Schaffner and R.S. Cohen (eds.): *PSA 1972.* Proceedings of the 3rd Biennial Meeting of the Philosophy of Science Association (Lansing, Michigan, Fall 1972). [Synthese Library 64] 1974
ISBN 90-277-0408-2; Pb 90-277-0409-0
21. R.S. Cohen and J.J. Stachel (eds.): *Selected Papers of Léon Rosenfeld.* [Synthese Library 100] 1979 ISBN 90-277-0651-4; Pb 90-277-0652-2
22. M. Čapek (ed.): *The Concepts of Space and Time.* Their Structure and Their Development. [Synthese Library 74] 1976
ISBN 90-277-0355-8; Pb 90-277-0375-2
23. M. Grene: *The Understanding of Nature.* Essays in the Philosophy of Biology. [Synthese Library 66] 1974 ISBN 90-277-0462-7; Pb 90-277-0463-5
24. D. Ihde: *Technics and Praxis.* A Philosophy of Technology. [Synthese Library 130] 1979 ISBN 90-277-0953-X; Pb 90-277-0954-8
25. J. Hintikka and U. Remes: *The Method of Analysis.* Its Geometrical Origin and Its General Significance. [Synthese Library 75] 1974
ISBN 90-277-0532-1; Pb 90-277-0543-7
26. J.E. Murdoch and E.D. Sylla (eds.): *The Cultural Context of Medieval Learning.* Proceedings of the First International Colloquium on Philosophy,

Science, and Theology in the Middle Ages, 1973. [Synthese Library 76] 1975
ISBN 90-277-0560-7; Pb 90-277-0587-9

27. M. Grene and E. Mendelsohn (eds.): *Topics in the Philosophy of Biology.* [Synthese Library 84] 1976 ISBN 90-277-0595-X; Pb 90-277-0596-8
28. J. Agassi: *Science in Flux.* [Synthese Library 80] 1975
ISBN 90-277-0584-4; Pb 90-277-0612-3
29. J.J. Wiatr (ed.): *Polish Essays in the Methodology of the Social Sciences.* [Synthese Library 131] 1979 ISBN 90-277-0723-5; Pb 90-277-0956-4
30. P. Janich: *Protophysics of Time.* Constructive Foundation and History of Time Measurement. Translated from the 2nd German edition. 1985
ISBN 90-277-0724-3
31. R.S. Cohen and M.W. Wartofsky (eds.): *Language, Logic, and Method.* 1983
ISBN 90-277-0725-1
32. R.S. Cohen, C.A. Hooker, A.C. Michalos and J.W. van Evra (eds.): *PSA 1974.* Proceedings of the 4th Biennial Meeting of the Philosophy of Science Association. [Synthese Library 101] 1976
ISBN 90-277-0647-6; Pb 90-277-0648-4
33. G. Holton and W.A. Blanpied (eds.): *Science and Its Public.* The Changing Relationship. [Synthese Library 96] 1976
ISBN 90-277-0657-3; Pb 90-277-0658-1
34. M.D. Grmek, R.S. Cohen and G. Cimino (eds.): *On Scientific Discovery.* The 1977 Erice Lectures. 1981 ISBN 90-277-1122-4; Pb 90-277-1123-2
35. S. Amsterdamski: *Between Expeience and Metaphysics.* Philosophical Problems of the Evolution of Science. Translated from Polish. [Synthese Library 77] 1975 ISBN 90-277-0568-2; Pb 90-277-0580-1
36. M. Marković and G. Petrović (eds.): *Praxis.* Yugoslav Essays in the Philosophy and Methodology of the Social Sciences. [Synthese Library 134] 1979
ISBN 90-277-0727-8; Pb 90-277-0968-8
37. H. von Helmholtz: *Epistemological Writings.* The Paul Hertz / Moritz Schlick Centenary Edition of 1921. Translated from German by M.F. Lowe. Edited with an Introduction and Bibliography by R.S. Cohen and Y. Elkana. [Synthese Library 79] 1977 ISBN 90-277-0290-X; Pb 90-277-0582-8
38. R.M. Martin: *Pragmatics, Truth and Language.* 1979
ISBN 90-277-0992-0; Pb 90-277-0993-9
39. R.S. Cohen, P.K. Feyerabend and M.W. Wartofsky (eds.): *Essays in Memory of Imre Lakatos.* [Synthese Library 99] 1976
ISBN 90-277-0654-9; Pb 90-277-0655-7
40. B.M Kedrov and V. Sadovsky (eds.): Current Soviet Studies in the Philosophy of Science. (In prep.) ISBN 90-277-0729-4
41. M. Raphael: Theorie des geistigen Schaffens aus marxistischer Grundlage. (In prep.) ISBN 90-277-0730-8

## Boston Studies in the Philosophy of Science

42. H.R. Maturana and F.J. Varela: *Autopoiesis and Cognition.* The Realization of the Living. With a Preface to 'Autopoiesis' by S. Beer. 1980
ISBN 90-277-1015-5; Pb 90-277-1016-3
43. A. Kasher (ed.): *Language in Focus: Foundations, Methods and Systems.* Essays in Memory of Yehoshua Bar-Hillel. [Synthese Library 89] 1976
ISBN 90-277-0644-1; Pb 90-277-0645-X
44. T.D. Thao: *Investigations into the Origin of Language and Consciousness.* 1984 ISBN 90-277-0827-4
45. A. Ishmimoto (ed.): Japanese Studies in the History and Philosophy of Science. (In prep.) ISBN 90-277-0733-3
46. P.L. Kapitza: *Experiment, Theory, Practice.* Articles and Addresses. Edited by R.S. Cohen. 1980 ISBN 90-277-1061-9; Pb 90-277-1062-7
47. M.L. Dalla Chiara (ed.): *Italian Studies in the Philosophy of Science.* 1981
ISBN 90-277-0735-9; Pb 90-277-1073-2
48. M.W. Wartofsky: *Models.* Representation and the Scientific Understanding. [Synthese Library 129] 1979 ISBN 90-277-0736-7; Pb 90-277-0947-5
49. T.D. Thao: *Phenomenology and Dialectical Materialism.* Edited by R.S. Cohen. 1986 ISBN 90-277-0737-5
50. Y. Fried and J. Agassi: *Paranoia.* A Study in Diagnosis. [Synthese Library 102] 1976 ISBN 90-277-0704-9; Pb 90-277-0705-7
51. K.H. Wolff: *Surrender and Cath.* Experience and Inquiry Today. [Synthese Library 105] 1976 ISBN 90-277-0758-8; Pb 90-277-0765-0
52. K. Kosík: *Dialectics of the Concrete.* A Study on Problems of Man and World. 1976 ISBN 90-277-0761-8; Pb 90-277-0764-2
53. N. Goodman: *The Structue of Appearance.* [Synthese Library] 1977
ISBN 90-277-0773-1; Pb 90-277-0774-X
54. H.A. Simon: *Models of Discovery* and Other Topics in the Methods of Science. [Synthese Library 114] 1977 ISBN 90-277-0812-6; Pb 90-277-0858-4
55. M. Lazerowitz: *The Language of Philosophy.* Freud and Wittgenstein. [Synthese Library 117] 1977 ISBN 90-277-0826-6; Pb 90-277-0862-2
56. T. Nickles (ed.): *Scientific Discovery, Logic, and Rationality.* 1980
ISBN 90-277-1069-4; Pb 90-277-1070-8
57. J. Margolis: *Persons and Mind.* The Prospects of Nonreductive Materialism. [Synthese Library 121] 1978 ISBN 90-277-0854-1; Pb 90-277-0863-0
58. G. Radnitzky and G. Andersson (eds.): *Progress and Rationality in Science.* [Synthese Library 125] 1978 ISBN 90-277-0921-1; Pb 90-277-0922-X
59. G. Radnitzky and G. Andersson (eds.): *The Structure and Development of Science.* [Synthese Library 136] 1979
ISBN 90-277-0994-7; Pb 90-277-0995-5
60. T. Nickles (ed.): *Scientific Discovery.* Case Studies. 1980
ISBN 90-277-1092-9; Pb 90-277-1093-7

## Boston Studies in the Philosophy of Science

61. M.A. Finocchiaro: *Galileo and the Art of Reasoning.* Rhetorical Foundation of Logic and Scientific Method. 1980 ISBN 90-277-1094-5; Pb 90-277-1095-3
62. W.A. Wallace: *Prelude to Galileo.* Essays on Medieval and 16th-Century Sources of Galileo's Thought. 1981 ISBN 90-277-1215-8; Pb 90-277-1216-6
63. F. Rapp: *Analytical Philosophy of Technology.* Translated from German. 1981 ISBN 90-277-1221-2; Pb 90-277-1222-0
64. R.S. Cohen and M.W. Wartofsky (eds.): *Hegel and the Sciences.* 1984 ISBN 90-277-0726-X
65. J. Agassi: *Science and Society.* Studies in the Sociology of Science. 1981 ISBN 90-277-1244-1; Pb 90-277-1245-X
66. L. Tondl: *Problems of Semantics.* A Contribution to the Analysis of the Language of Science. Translated from Czech. 1981 ISBN 90-277-0148-2; Pb 90-277-0316-7
67. J. Agassi and R.S. Cohen (eds.): *Scientific Philosophy Today.* Essays in Honor of Mario Bunge. 1982 ISBN 90-277-1262-X; Pb 90-277-1263-8
68. W. Krajewski (ed.): *Polish Essays in the Philosophy of the Natural Sciences.* Translated from Polish and edited by R.S. Cohen and C.R. Fawcett. 1982 ISBN 90-277-1286-7; Pb 90-277-1287-5
69. J.H. Fetzer: *Scientific Knowledge.* Causation, Explanation and Corroboration. 1981 ISBN 90-277-1335-9; Pb 90-277-1336-7
70. S. Grossberg: *Studies of Mind and Brain.* Neural Principles of Learning, Perception, Development, Cognition, and Motor Control. 1982 ISBN 90-277-1359-6; Pb 90-277-1360-X
71. R.S. Cohen and M.W. Wartofsky (eds.): *Epistemology, Methodology, and the Social Sciences.* 1983. ISBN 90-277-1454-1
72. K. Berka: *Measurement.* Its Concepts, Theories and Problems. Translated from Czech. 1983 ISBN 90-277-1416-9
73. G.L. Pandit: *The Structure and Growth of Scientific Knowledge.* A Study in the Methodology of Epistemic Appraisal. 1983 ISBN 90-277-1434-7
74. A.A. Zinov'ev: *Logical Physics.* Translated from Russian. Edited by R.S. Cohen. 1983 ISBN 90-277-0734-0
    *See also* Volume 9.
75. G-G. Granger: *Formal Thought and the Sciences of Man.* Translated from French. With and Introduction by A. Rosenberg. 1983 ISBN 90-277-1524-6
76. R.S. Cohen and L. Laudan (eds.): *Physics, Philosophy and Psychoanalysis.* Essays in Honor of Adolf Grünbaum. 1983 ISBN 90-277-1533-5
77. G. Böhme, W. van den Daele, R. Hohlfeld, W. Krohn and W. Schäfer: *Finalization in Science.* The Social Orientation of Scientific Progress. Translated from German. Edited by W. Schäfer. 1983 ISBN 90-277-1549-1
78. D. Shapere: *Reason and the Search for Knowledge.* Investigations in the Philosophy of Science. 1984 ISBN 90-277-1551-3; Pb 90-277-1641-2

# Boston Studies in the Philosophy of Science

79. G. Andersson (ed.): *Rationality in Science and Politics*. Translated from German. 1984 ISBN 90-277-1575-0; Pb 90-277-1953-5
80. P.T. Durbin and F. Rapp (eds.): *Philosophy and Technology*. [*Also* Philosophy and Technology Series, Vol. 1] 1983 ISBN 90-277-1576-9
81. M. Marković: *Dialectical Theory of Meaning*. Translated from Serbo-Croat. 1984 ISBN 90-277-1596-3
82. R.S. Cohen and M.W. Wartofsky (eds.): *Physical Sciences and History of Physics*. 1984. ISBN 90-277-1615-3
83. É. Meyerson: *The Relativistic Deduction*. Epistemological Implications of the Theory of Relativity. Translated from French. With a Review by Albert Einstein and an Introduction by Milić Čapek. 1985 ISBN 90-277-1699-4
84. R.S. Cohen and M.W. Wartofsky (eds.): *Methodology, Metaphysics and the History of Science*. In Memory of Benjamin Nelson. 1984 ISBN 90-277-1711-7
85. G. Tamás: *The Logic of Categories*. Translated from Hungarian. Edited by R.S. Cohen. 1986 ISBN 90-277-1742-7
86. S.L. De C. Fernandes: *Foundations of Objective Knowledge*. The Relations of Popper's Theory of Knowledge to That of Kant. 1985 ISBN 90-277-1809-1
87. R.S. Cohen and T. Schnelle (eds.): *Cognition and Fact*. Materials on Ludwik Fleck. 1986 ISBN 90-277-1902-0
88. G. Freudenthal: *Atom and Individual in the Age of Newton*. On the Genesis of the Mechanistic World View. Translated from German. 1986 ISBN 90-277-1905-5
89. A. Donagan, A.N. Perovich Jr and M.V. Wedin (eds.): *Human Nature and Natural Knowledge*. Essays presented to Majorie Grene on the Occasion of Her 75th Birthday. 1986 ISBN 90-277-1974-8
90. C. Mitcham and A. Hunning (eds.): *Philosophy and Technology II*. Information Technology and Computers in Theory and Practice. [*Also* Philosophy and Technology Series, Vol. 2] 1986 ISBN 90-277-1975-6
91. M. Grene and D. Nails (eds.): *Spinoza and the Sciences*. 1986 ISBN 90-277-1976-4
92. S.P. Turner: *The Search for a Methodology of Social Science*. Durkheim, Weber, and the 19th-Century Problem of Cause, Probability, and Action. 1986. ISBN 90-277-2067-3
93. I.C. Jarvie: *Thinking about Society*. Theory and Practice. 1986 ISBN 90-277-2068-1
94. E. Ullmann-Margalit (ed.): *The Kaleidoscope of Science*. The Israel Colloquium: Studies in History, Philosophy, and Sociology of Science, Vol. I. 1986 ISBN 90-277-2158-0; Pb 90-277-2159-9
95. E. Ullmann-Margalit (ed.): *The Prism of Science*. The Israel Colloquium: Studies in History, Philosophy, and Sociology of Science, Vol. II. 1986 ISBN 90-277-2160-2; Pb 90-277-2161-0

## Boston Studies in the Philosophy of Science

96. G. Márkus: *Language and Production.* A Critique of the Paradigms. Translated from French. 1986 ISBN 90-277-2169-6
97. F. Amrine, F.J. Zucker and H. Wheeler (eds.): *Goethe and the Sciences: A Reappraisal.* 1987 ISBN 90-277-2265-X; Pb 90-277-2400-8
98. J.C. Pitt and M. Pera (eds.): *Rational Changes in Science.* Essays on Scientific Reasoning. Translated from Italian. 1987 ISBN 90-277-2417-2
99. O. Costa de Beauregard: *Time, the Physical Magnitude.* 1987 ISBN 90-277-2444-X
100. A. Shimony and D. Nails (eds.): *Naturalistic Epistemology.* A Symposium of Two Decades. 1987 ISBN 90-277-2337-0
101. N. Rotenstreich: *Time and Meaning in History.* 1987 ISBN 90-277-2467-9
102. D.B. Zilberman: *The Birth of Meaning in Hindu Thought.* Edited by R.S. Cohen. 1988 ISBN 90-277-2497-0
103. T.F. Glick (ed.): *The Comparative Reception of Relativity.* 1987 ISBN 90-277-2498-9
104. Z. Harris, M. Gottfried, T. Ryckman, P. Mattick Jr, A. Daladier, T.N. Harris and S. Harris: *The Form of Information in Science.* Analysis of an Immunology Sublanguage. With a Preface by Hilary Putnam. 1989 ISBN 90-277-2516-0
105. F. Burwick (ed.): *Approaches to Organic Form.* Permutations in Science and Culture. 1987 ISBN 90-277-2541-1
106. M. Almási: *The Philosophy of Appearances.* Translated from Hungarian. 1989 ISBN 90-277-2150-5
107. S. Hook, W.L. O'Neill and R. O'Toole (eds.): *Philosophy, History and Social Action.* Essays in Honor of Lewis Feuer. With an Autobiographical Essay by L. Feuer. 1988 ISBN 90-277-2644-2
108. I. Hronszky, M. Fehér and B. Dajka: *Scientific Knowledge Socialized.* Selected Proceedings of the 5th Joint International Conference on the History and Philosophy of Science organized by the IUHPS (Veszprém, Hungary, 1984). 1988 ISBN 90-277-2284-6
109. P. Tillers and E.D. Green (eds.): *Probability and Inference in the Law of Evidence.* The Uses and Limits of Bayesianism. 1988 ISBN 90-277-2689-2
110. E. Ullmann-Margalit (ed.): *Science in Reflection.* The Israel Colloquium: Studies in History, Philosophy, and Sociology of Science, Vol. III. 1988 ISBN 90-277-2712-0; Pb 90-277-2713-9
    *See also* Volumes 94 and 95.
111. K. Gavroglu, Y. Goudaroulis and P. Nicolacopoulos (eds.): *Imre Lakatos and Theories of Scientific Change.* 1989 ISBN 90-277-2766-X
112. B. Glassner and J.D. Moreno (eds.): *The Qualitative- Quantitative Distinction in the Social Sciences.* 1989 ISBN 90-277-2829-1
113. K. Arens: *Structures of Knowing.* Psychologies of the 19th Century. 1989 ISBN 0-7923-0009-2

## Boston Studies in the Philosophy of Science

114. A. Janik: *Style, Politics and the Future of Philosophy.* 1989
ISBN 0-7923-0056-4
115. F. Amrine (ed.): *Literature and Science as Modes of Expression.* With an Introduction by S. Weininger. 1989 ISBN 0-7923-0133-1
116. J.R. Brown and J. Mittelstrass (eds.): *An Intimate Relation.* Studies in the History and Philosophy of Science. Presented to Robert E. Butts on His 60th Birthday. 1989 ISBN 0-7923-0169-2
117. F. D'Agostino and I.C. Jarvie (eds.): *Freedom and Rationality.* Essays in Honor of John Watkins. 1989 ISBN 0-7923-0264-8
118. D. Zolo: *Reflexive Epistemology.* The Philosophical Legacy of Otto Neurath. 1989 ISBN 0-7923-0320-2
119. M. Kearn, B.S. Philips and R.S. Cohen (eds.): *Georg Simmel and Contemporary Sociology.* 1989 ISBN 0-7923-0407-1
120. T.H. Levere and W.R. Shea (eds.): *Nature, Experiment and the Science.* Essays on Galileo and the Nature of Science. In Honour of Stillman Drake. 1989
ISBN 0-7923-0420-9
121. P. Nicolacopoulos (ed.): *Greek Studies in the Philosophy and History of Science.* 1990 ISBN 0-7923-0717-8
122. R. Cooke and D. Costantini (eds.): *Statistics in Science.* The Foundations of Statistical Methods in Biology, Physics and Economics. 1990
ISBN 0-7923-0797-6
123. P. Duhem: *The Origins of Statics.* Translated from French by G.F. Leneaux, V.N. Vagliente and G.H. Wagner. With an Introduction by S.L. Jaki. 1991
ISBN 0-7923-0898-0
124. H. Kamerlingh Onnes: *Through Measurement to Knowledge.* The Selected Papers, 1853-1926. Edited and with an Introduction by K. Gavroglu and Y. Goudaroulis. 1991 ISBN 0-7923-0825-5
125. M. Čapek: *The New Aspects of Time: Its Continuity and Novelties.* Selected Papers in the Philosophy of Science. 1991 ISBN 0-7923-0911-1
126. S. Unguru (ed.): *Physics, Cosmology and Astronomy, 1300- 1700.* Tension and Accomodation. 1991 ISBN 0-7923-1022-5
127. Z. Bechler: *Newton's Physics on the Conceptual Structure of the Scientific Revolution.* 1991 ISBN 0-7923-1054-3
128. É. Meyerson: *Explanation in the Sciences.* Translated from French by M-A. Siple and D.A. Siple. 1991 ISBN 0-7923-1129-9
129. A.I. Tauber (ed.): *Organism and the Origins of Self.* 1991
ISBN 0-7923-1185-X
130. F.J. Varela and J-P. Dupuy (eds.): *Understanding Origins.* Contemporary Views on the Origin of Life, Mind and Society. 1992 ISBN 0-7923-1251-1
131. G.L. Pandit: *Methodological Variance.* Essays in Epistemological Ontology and the Methodology of Science. 1991 ISBN 0-7923-1263-5

## Boston Studies in the Philosophy of Science

132. G. Munévar (ed.): *Beyond Reason.* Essays on the Philosophy of Paul Feyerabend. 1991 ISBN 0-7923-1272-4
133. T.E. Uebel (ed.): *Rediscovering the Forgotten Vienne Circle.* Austrian Studies on Otto Neurath and the Vienna Circle. Partly translated from German. 1991 ISBN 0-7923-1276-7
134. W.R. Woodward and R.S. Cohen (eds.): *World Views and Scientific Discipline Formation.* Science Studies in the [former] German Democratic Republic. Partly translated from German by W.R. Woodward. 1991 ISBN 0-7923-1286-4
135. P. Zambelli: *The Speculum Astronomiae and Its Enigma.* Astrology, Theology and Science at Albertus Magnus' Time. (forthcoming) ISBN 0-7923-1380-1
136. P. Petitjean, C. Jami and A.M. Moulin (eds.): *Science and Empires.* Historical Studies about Scientific Development and European Expansion. ISBN 0-7923-1518-9

*Also of interest:*

R.S. Cohen and M.W. Wartofsky (eds.): *A Portrait of Twenty-Five Years Boston Colloquia for the Philosophy of Science, 1960-1985.* 1985 ISBN Pb 90-277-1971-3

*Previous volumes are still available.*

KLUWER ACADEMIC PUBLISHERS – DORDRECHT / BOSTON / LONDON